关于鞅的一些主题和模型

Some Topics and Models Related to Martingales

郝顺利　著
Shunli Hao

北京第二外国语学院基础科学部　北京，中国
Department of Basic Sciences, Beijing International Studies University, Beijing, China

旅游教育出版社
·北京·

策　　划：何　玲

责任编辑：何　玲

图书在版编目（ＣＩＰ）数据

关于鞅的一些主题和模型 ／ 郝顺利著. -- 北京：
旅游教育出版社，2021.1

ISBN 978-7-5637-4204-2

Ⅰ. ①关… Ⅱ. ①郝… Ⅲ. ①鞅－研究 Ⅳ.
①O211.6

中国版本图书馆CIP数据核字 (2020) 第267918号

关于鞅的一些主题和模型

Guanyu Yang de Yixie Zhuti he Moxing

郝顺利　著

出版单位	旅游教育出版社
地　　址	北京市朝阳区定福庄南里 1 号
邮　　编	100024
发行电话	（010）65778403　65728372　65767462（传真）
本社网址	www.tepcb.com
E - mail	tepfx@163.com
印刷单位	北京玺诚印务有限公司
经销单位	新华书店
开　　本	710毫米×1000毫米　1/16
印　　张	9
字　　数	130 千字
版　　次	2021 年 1 月第 1 版
印　　次	2021 年 1 月第 1 次印刷
定　　价	60.00 元

（图书如有装订差错请与发行部联系）

Preface

The major purpose of this book is to study some specific topics and models related to certain martingales. The book contains six chapters. Our main contribution here is to provide some new research directions and probabilistic methods on some topics and models related to martingales for the readers.

In Chapter 0, we introduce some basic concepts of probability theory and their elementary properties which can be found in some classic books, we also introduce some specific probabilistic models which can be found in some new papers. These concepts and models include probability spaces, random variables, random elements, expectation, conditional expectation, conditional probability, convergence, weak convergence, martingale, slowly varing function, the laws of large numbers, the central limit theorems, branching process in a random environment, multiplicative cascades in a random environment, directed polymers in a random environment, etc.

In Chapter 1, we study the complete convergence for double-indexed randomly weighted sums of the triangular array of Banach space valued martingale differences. We give a sufficient condition for the complete convergence of double-indexed randomly weighted sums of the triangular array $\{(X_{nj}, \mathcal{F}_{nj}), 1 \leq j \leq n, n \geq 1\}$ of Banach space valued martingale differences and extend the corresponding result of Dedecker and Merlevède (2008) from partial sums of the sequence of Banach space valued martingale differences to double-indexed randomly weighted ones of the triangular array of Banach space valued martingale differences.

In Chapter 2, we study mainly the complete moment convergence for double-indexed randomly weighted sums of the triangular array of Banach space valued martingale differences. We give some sufficient conditions for the complete moment convergence of double-indexed randomly weighted sums of the triangular array $\{(X_{nj}, \mathcal{F}_{nj}), 1 \leq j \leq n, n \geq 1\}$ of Banach space valued martingale differences and extend the corresponding results of Yang $et\ al.$ (2014), Wang and Hu (2014), Yang $et\ al.$ (2013) and Wang $et\ al.$ (2012) to double-indexed randomly weighted sums of the triangular array of Banach space valued martingale differences.

Chapter 3 is concerned with local dimensions of the branching measure in a random environment. Let $\zeta = (\zeta_0,\ \zeta_1,\ \cdots)$ be a sequence of independent and identically distributed random variables, taking values in some space Θ. Let $\mu = \mu_\omega$ be the branching measure on the boundary $\partial\mathbb{T}$ of the supercritical Galton-Watson tree $\mathbb{T} = \mathbb{T}(\omega)$ of branching precess in the random environment ζ. Denote by $\underline{d}(\mu, u)$ and $\overline{d}(\mu, u)$ the lower and upper local dimensions of μ at $u \in \partial\mathbb{T}$, respectively. We find a sufficient condition under which almost surely ζ, $\mathbb{P}_\zeta$-almost surely $\underline{d}(\mu, u) = \overline{d}(\mu, u) = \mathbb{E}\log m_0$ for all $u \in \partial\mathbb{T}$, where m_0 is the expected value of the offspring distribution for an initial partical. The result extends that of Liu (2001).

In Chapter 4, we study the limit theorems for multiplicative cascades in a random environment. Let $\zeta = (\zeta_0,\ \zeta_1,\ \cdots)$ be a sequence of independent and identically distributed random variables. For $r \geq 2$, let μ_r be Mandelbrot's (limit) measure of multiplicative cascades defined with positive weights indexed by nodes of a regular r-ary tree, and let $Z^{(r)}$ be the mass of μ_r. We study asymptotic properties of $Z^{(r)}$ and the sequence of random measures $(\mu_r)_r$ as $r \to \infty$. We obtain some laws of large numbers and a central limit theorem. The results extend ones established by Liu and Rouault (2000) and by Liu, Rio and Rouault (2003).

In Chapter 5, we study the convergence of the free energy of directed polymers in a random environment. We establish a new concentration inequality for the free energy of directed polymers in a random environment and obtain the result that the centered energy converges to 0 in $\mathbb{L}^p$ for $p > 2d$ (d is the dimension of integer lattice) assuming only the existence of the p-th moment of the environment variable. The convergence result is supplementary to some known results.

I wish to thank the support of the project of the school level publishing fund of Beijing International Studies University (北京第二外国语学院校级出版基金资助项目), which enabled me to do research for past 1 year and complete the writing of the book at Beijing International Studies University. Thank my colleagues in Beijing International Studies University for their help. I would like to acknowledge my great debt to previous works on this subject by Zhiqiang Gao (高志强) and Xingang Liang (梁新刚) . Last but not the least, I give thanks to my supervisor, Quansheng Liu (刘全升) , for his patience and support throughout the task.

Shunli Hao
Beijing, China 2020

The work is supported by the project of the school level publishing fund of Beijing International Studies University.

Contents

Chapter 0

Preliminary knowledge

In this chapter, we introduce some basic concepts of probability theory and their elementary properties, we also introduce some probabilistic models. We can find them in some classical books (see, for example, [74, 17]) and new papers.

0.1 Probability spaces and random variables (elements)

Definition 0.1 *Let Ω be a nonempty set. Let $\mathcal{F}$ be a σ-field of subsets of Ω, that is, a nonempty class of subsets of Ω which contains Ω and is closed under countable union and complementation. Let $\mathbb{P}$ be a measure defined on $\mathcal{F}$ satisfying $\mathbb{P}(\Omega) = 1$. Then the triple $(\Omega, \mathcal{F}, \mathbb{P})$ is called a probability space, and $\mathbb{P}$, a probability measure.*

The set Ω is the sure event, and elements of $\mathcal{F}$ are called events. Singleton sets $\{\omega\}$ are called elementary events. The symbol $\emptyset$ denotes the empty set and is known as the null or impossible event. Unless otherwise stated, the probability space $(\Omega, \mathcal{F}, \mathbb{P})$ is fixed, and A, B, C, $\ldots$, with or without subscripts, represent events.

We note that, if $A_n \in \mathcal{F}$, $n = 1, 2, \ldots$, then A_n^c, $\bigcup_{n=1}^{\infty} A_n$, $\bigcap_{n=1}^{\infty} A_n$, $\liminf_{n \to \infty} A_n$, $\limsup_{n \to \infty} A_n$ and $\lim_{n \to \infty} A_n$ (if it exists) are events. Also, the probability measure $\mathbb{P}$ is defined on $\mathcal{F}$, and for all events A, A_n,

$$\mathbb{P}(A) \geq 0, \quad \mathbb{P}\left(\bigcup_{n=1}^{\infty} A_n\right) = \sum_{n=1}^{\infty} \mathbb{P}(A_n) \ (A_n\text{'s disjoint}), \quad \mathbb{P}(\Omega) = 1.$$

It follows that

$$\mathbb{P}(\emptyset) = 0, \quad \mathbb{P}(A) \geq \mathbb{P}(B) \quad \text{for } A \subset B, \quad \mathbb{P}\left(\bigcup_{n=1}^{\infty} A_n\right) \leq \sum_{n=1}^{\infty} \mathbb{P}(A_n).$$

Moreover,

$$\mathbb{P}\left(\liminf_{n\to\infty} A_n\right) \leq \liminf_{n\to\infty} \mathbb{P}(A_n) \leq \limsup_{n\to\infty} \mathbb{P}(A_n) \leq \mathbb{P}\left(\limsup_{n\to\infty} A_n\right),$$

and, if $\lim_{n\to\infty} A_n$ exists, then

$$\mathbb{P}\left(\lim_{n\to\infty} A_n\right) = \lim_{n\to\infty} \mathbb{P}(A_n).$$

The last result is known as the continuity property of probability measures.

Example 0.1 *Let $\Omega = \{\omega_j : j \geq 1\}$, and let $\mathcal{F}$ be the σ-field of all subsets of Ω. Let $\{p_j, j \geq 1\}$ be any sequence of nonnegative real numbers satisfying $\sum_{j=1}^{\infty} p_j = 1$. Define $\mathbb{P}$ on $\mathcal{F}$ by*

$$\mathbb{P}(E) = \sum_{\omega_j \in E} p_j, \quad E \in \mathcal{F}.$$

Then $\mathbb{P}$ defines a probability measure on $(\Omega, \mathcal{F})$, and $(\Omega, \mathcal{F}, \mathbb{P})$ is a probability space.

Example 0.2 *Let $\Omega = (0, 1]$ and $\mathcal{F} = \mathcal{B}$ be the σ-field of Borel sets on Ω. Let λ be the Lebesgue measure on $\mathcal{B}$. Then $(\Omega, \mathcal{F}, \lambda)$ is a probability space.*

Definition 0.2 *Let $(\Omega, \mathcal{F}, \mathbb{P})$ be a probability space. A real-valued function X defined on Ω is said to be a random variable if*

$$X^{-1}(E) = \{\omega \in \Omega : X(\omega) \in E\} \in \mathcal{F} \quad \text{for all } E \in \mathcal{B}_1$$

where $\mathcal{B}_1$ is the σ-field of Borel sets in $\mathbb{R} = (-\infty, \infty)$; that is, a random variable X is a measurable transformation of $(\Omega, \mathcal{F}, \mathbb{P})$ into $(\mathbb{R}, \mathcal{B}_1)$.

We note that it suffices to require that $X^{-1}(I) \in \mathcal{F}$ for all intervals I in $\mathbb{R}$, or for all semiclosed intervals $I = (a, b]$, or for all intervals $I = (-\infty, b]$, and so on. Unless otherwise specified, X, Y, $\ldots$, with or without subscripts, will represent random variables.

We note that a random variable X defined on $(\Omega, \mathcal{F}, \mathbb{P})$ induces a measure $\mathbb{P}_X$ on $\mathcal{B}_1$ defined by the relation

$$\mathbb{P}_X(E) = \mathbb{P}\{X^{-1}(E)\} \quad (E \in \mathcal{B}_1).$$

Clearly $\mathbb{P}_X$ is a probability measure on $\mathcal{B}_1$ and is called the probability distribution or, simply, the distribution of X. We note that $\mathbb{P}_X$ is a Lebesgue-Stieltjes measure on $\mathcal{B}_1$.

Definition 0.3 *For every $x \in \mathbb{R}$ set*

$$F_X(x) = \mathbb{P}_X(-\infty, x] = \mathbb{P}\{\omega \in \Omega : \ X(\omega) \le x\}. \tag{0.1}$$

We call $F_X = F$ the distribution function of the random variable X.

The following theorem is an elementary property of a distribution function.

Theorem 0.1 *The distribution function F of a random variable X is a nondecreasing, right-continuous function on $\mathbb{R}$ which satisfies*

$$F(-\infty) = \lim_{x \to -\infty} F(x) = 0$$

and

$$F(+\infty) = \lim_{x \to +\infty} F(x) = 1.$$

Corollary 0.1 *A distribution function F is continuous at $x \in \mathbb{R}$ if and only if $\mathbb{P}\{\omega : X(\omega) = x\} = 0$.*

Remark 0.2 *Let X be a random variable, and let g be a Borel-measurable function defined on $\mathbb{R}$. Then $g(X)$ is also a random variable whose distribution is determined by that of X.*

The following theorem show that a function F on $\mathbb{R}$ with the properties stated in Theorem 0.1 determines uniquely a probability measure $\mathbb{P}_F$ on $\mathcal{B}_1$.

Theorem 0.3 *Let F be a nondecreasing, right-continuous function defined on $\mathbb{R}$ and satisfying*

$$F(-\infty) = 0$$

and

$$F(+\infty) = 1.$$

Then there exists a probability measure $\mathbb{P} = \mathbb{P}_F$ on $\mathcal{B}_1$ determined uniquely by the relation

$$\mathbb{P}_F(-\infty, \ x] = F(x)$$

for every $x \in \mathbb{R}$.

Remark 0.4 *Let F be a bounded nondecreasing, right-continuous function defined on $\mathbb{R}$ satisfying $F(-\infty) = 0$. Then there exists a finite measure $\mu = \mu_F$ on $\mathcal{B}_1$ determined uniquely by the relation*

$$\mu_F(-\infty, \ x] = F(x)$$

for every $x \in \mathbb{R}$.

Remark 0.5 *Let F on $\mathbb{R}$ satisfy the conditions of Theorem 0.3. Then there exists a random variable X on some probability space such that F is the distribution function of X. In fact, consider the probability space $(\Omega, \mathcal{B}, \mathbb{P})$, where $\mathbb{P}$ is the probability measure as constructed in Theorem 0.3. Let $X(\omega) = \omega$, for all $\omega \in \mathbb{R}$. It is easy to see that F is the distribution function of the random variable X.*

Let F be a distribution function, and let $x \in \mathbb{R}$ be a discontinuity point of F. Then $p(x) = F(x) - F(x - 0)$ is called the jump of F at x. A point is said to be a point of increase of F if, for every $\varepsilon > 0$, $F(x + \varepsilon) - F(x - \varepsilon) > 0$.

Some other elementary properties of a distribution function are obtained in the following propositions.

Proposition 0.1 *Let F_1 and F_2 be two distribution functions such that*

$$F_1(x) = F_2(x) \quad \text{for all } x \in D,$$

where D is everywhere dense in $\mathbb{R}$. Then $F_1(x) = F_2(x)$ for every $x \in \mathbb{R}$.

Proposition 0.2 *The set of discontinuity points of a distribution function F is countable.*

Proposition 0.3 *Let F_1 and F_2 be two distribution functions, and let C_1 and C_2, respectively, be their sets of continuity points. If*

$$F_1(x) = F_2(x) \quad \text{for } x \in C_1 \cap C_2,$$

then $F_1(x) = F_2(x)$ for all $x \in \mathbb{R}$.

Let $(\Omega, \mathcal{F}, \mathbb{P})$ be a probability space, and $\mathbb{B}$ be a real Banach space with norm $\|\cdot\|$. Clearly $\mathbb{B}$ is a topological vector space with respect to the metric topology induced by $\|\cdot\|$. Let $\mathcal{B}_2$ be the σ-field generated by the class of all open subsets of $\mathbb{B}$. Then $\mathcal{B}_2$ is known as the Borel σ-field on $\mathbb{B}$, and elements of $\mathcal{B}_2$ are called Borel sets.

Definition 0.4 *A mapping $X : \Omega \to \mathbb{B}$ is called a $\mathbb{B}$-valued random element if X is $\mathcal{B}_2$-measurable, that is, for every $E \in \mathcal{B}_2$,*

$$X^{-1}(E) = \{\omega \in \Omega : X(\omega) \in E\} \in \mathcal{F}.$$

Remark 0.6 *Let $X : (\Omega, \mathcal{F}, \mathbb{P}) \to (\mathbb{B}, \mathcal{B}_2)$ be a $\mathbb{B}$-valued random element, and let $\mathbb{B}_0$ be another Banach space with Borel σ-field $\mathcal{B}_0$. Let T be a measurable mapping of $(\mathbb{B}, \mathcal{B}_2)$ into $(\mathbb{B}_0, \mathcal{B}_0)$. Then $T(X)$ is a $\mathbb{B}_0$-valued random element.*

Remark 0.7 *Let $\mathbb{B}^*$ be the dual (or conjugate) of $\mathbb{B}$, that is, $\mathbb{B}^*$ is the Banach space consisting of all bounded (continuous) linear functional on $\mathbb{B}$. Then it follows immediately from Remark 0.6 that, for every $l \in \mathbb{B}^*$, $l(X)$ is a real-valued random variable. In particular, if $\mathbb{B}$ is separable and $l(X)$ is a random variable for every $l \in \mathbb{B}^*$, then X is a $\mathbb{B}$-valued random element.*

Proposition 0.4 *Let* $\mathbb{B}$ *be a separable Banach space, and* X *a* $\mathbb{B}$-*valued random element. Then* $\|X\|$ *is a random variable.*

Let $\mathbb{P}_X$ be the set function on $\mathcal{B}_2$ defined by

$$\mathbb{P}_X(E) = \mathbb{P}(X^{-1}(E)), \quad E \in \mathcal{B}_2.$$

Clearly $\mathbb{P}_X$ is a probability measure on $\mathcal{B}_2$ induced by X and is known as the probablity distribution of X.

Definition 0.5 *Let* X *and* Y *be two* $\mathbb{B}$-*valued random elements defined on* $(\Omega, \mathcal{F}, \mathbb{P})$. *Then* X *and* Y *are said to be identically distributed if* $\mathbb{P}_X = \mathbb{P}_Y$. *A collection of* $\mathbb{B}$-*valued random elements is said to be identically distributed if every pair has the same probability distribution. In particular,* X *is said to be symmetric if* X *and* $-X$ *are identically distributed. Here* $-X$ *is the* $\mathbb{B}$-*valued random element defined by* $(-X)(\omega) = -X(\omega)$ *for all* $\omega \in \Omega$.

Remark 0.8 *Let* $\{X_\alpha, \alpha \in A\}$ *be a collection of identically distributed* $\mathbb{B}$-*valued random elements, and let* T *be a measurable mapping of* $\mathbb{B} \to \mathbb{B}_0$, *where* $\mathbb{B}_0$ *is a Banach space. Then* $\{T(X_\alpha), \alpha \in A\}$ *is a collection of identically distributed* $\mathbb{B}_0$-*valued random elements. In particular, if* $\mathbb{B}$ *is separable and, for every* $l \in \mathbb{B}^*$, $\{l(X_\alpha), \alpha \in A\}$ *is a collection of identically distributed random variables, then* $\{X_\alpha, \alpha \in A\}$ *are identically distributed.*

0.2 Expectation and conditional expectation

We now study some characteristics of a random variable (or of its distribution function). These play an important role in the study of probability theory.

Let $(\Omega, \mathcal{F}, \mathbb{P})$ be a probability space, and X be a random variable defined on it. Let g be a real-valued Borel-measurable function on $\mathbb{R}$. Then $g(X)$ is also a random variable.

Definition 0.6 *We say that the mathematical expectation (or, simply, the expectation) of* $g(X)$ *exists if* $g(X)$ *is integrable over* Ω *with respect to* $\mathbb{P}$. *In this case we define the expectation* $\mathbb{E}g(X)$ *of the random variable* $g(X)$ *by*

$$\mathbb{E}g(X) = \int_\Omega g(X(\omega))\mathrm{d}\mathbb{P}(\omega) = \int_\Omega g(X)\mathrm{d}\mathbb{P}.$$

Remark 0.9 *Since integrability is equivalent to absolute integrability, it follows that* $\mathbb{E}g(X)$ *exists if and only if* $\mathbb{E}|g(X)|$ *exists.*

Let $\mathbb{P}_X$ be the probability distribution of X. Suppose that $\mathbb{E}g(X)$ exists. Then it follows that g is also integrable over $\mathbb{R}$ with respect to $\mathbb{P}_X$. Moreover, the relation

$$\int_\Omega g(X)\mathrm{d}\mathbb{P} = \int_\mathbb{R} g(t)\mathrm{d}\mathbb{P}_X(t) \tag{0.2}$$

holds. We note that the integral on the right hand side of (0.2) is the Lebesgue-Stieltjes integral of g with respect to $\mathbb{P}_X$.

In particular, if g is continuous on $\mathbb{R}$ and $\mathbb{E}g(X)$ exists, we can rewrite (0.2) as follows:

$$\int_\Omega g(X)\mathrm{d}\mathbb{P} = \int_{\mathbb{R}} g\mathrm{d}\mathbb{P}_X = \int_{-\infty}^{\infty} g(x)\mathrm{d}F(x), \qquad (0.3)$$

where F is the distribution function corresponding to $\mathbb{P}_X$, and the last integral is a Riemann-Stieltjes integral. Two important special cases of (0.3) are as follows.

CASE 1. Let F be discrete with the set of discontinuity points $\{x_n,\ n = 1,\ 2,\ \ldots\}$. Let $p(x_n)$ be the jump of F at x_n, $n = 1,\ 2,\ \ldots$. Then $\mathbb{E}g(X)$ exists if and only if $\sum_{n=1}^{\infty} |g(x_n)|p(x_n) < \infty$, and in that case we have

$$\mathbb{E}g(X) = \sum_{n=1}^{\infty} g(x_n)p(x_n).$$

CASE 2. Let F be absolutely continuous on $\mathbb{R}$ with probability density function $f(x) = F'(x)$. Then $\mathbb{E}g(X)$ exists if and only if $\int_{-\infty}^{\infty} |g(x)|f(x)\mathrm{d}x < \infty$, and in that case we have

$$\mathbb{E}g(X) = \int_{-\infty}^{\infty} g(x)f(x)\mathrm{d}x.$$

We now state some elementary properties of random variables with finite expectations which follows as immediate consequences of the properties of integrable functions. Denote by $\mathbb{L}^1 = \mathbb{L}^1(\Omega, \mathcal{F}, \mathbb{P})$ the set of all random variables X on Ω such that $\mathbb{E}|X| < \infty$. In the following we write a.s. to abbreviate "almost surely (everywhere) with respect to the probability distribution of X on $(\mathbb{R}, \mathcal{B})$."
(a) $X, Y \in \mathbb{L}^1$ and $\alpha, \beta \in \mathbb{R} \Rightarrow \alpha X + \beta Y \in \mathbb{L}^1$ and $\mathbb{E}(\alpha X + \beta Y) = \alpha\mathbb{E}X + \beta\mathbb{E}Y$.
(b) $X \in \mathbb{L}^1 \Rightarrow |\mathbb{E}X| \leq \mathbb{E}|X|$.
(c) $X \in \mathbb{L}^1$, $X \geq 0$ a.s. $\Rightarrow \mathbb{E}X \geq 0$.
(d) Let $X \in \mathbb{L}^1$. Then $\mathbb{E}|X| = 0 \Leftrightarrow X = 0$ a.s.
(e) For $E \in \mathcal{F}$, write $\mathbf{1}_E$ for the indicator function of the set E, that is, $\mathbf{1}_E = 1$ on E and $= 0$ otherwise. Then $X \in \mathbb{L}^1 \Rightarrow X\mathbf{1}_E \in \mathbb{L}^1$, and we write

$$\int_E X\mathrm{d}\mathbb{P} = \mathbb{E}(X\mathbf{1}_E)$$

Also, $\mathbb{E}(|X|\mathbf{1}_E) = 0 \Leftrightarrow$ either $\mathbb{P}(E) = 0$ or $X = 0$ a.s. on E.
(f) If $X \in \mathbb{L}^1$, then $X = 0$ a.s. $\Leftrightarrow \mathbb{E}(X\mathbf{1}_E) = 0$ for all $E \in \mathcal{F}$.
(g) Let $X \in \mathbb{L}^1$, and define the set function Q_X on $\mathcal{F}$ by

$$Q_X(E) = \int_E X\mathrm{d}\mathbb{P}, \quad E \in \mathcal{F}$$

Then Q_X is countably additive on $\mathcal{F}$, and Q_X is absolutely continuous with respect to $\mathbb{P}$. In particular, Q_X is a finite measure on $\mathcal{F}$ if $X \geq 0$ a.s.

(h) Let $X \in \mathbb{L}^1$ and $E \in \mathcal{F}$. If $\alpha \leq X \leq \beta$ a.s. on E for α, $\beta \in \mathbb{R}$, then

$$\alpha \mathbb{P}(E) \leq \int_E X d\mathbb{P} \leq \beta \mathbb{P}(E).$$

(i) Let $Y \in \mathbb{L}^1$ and X be a random variable such that $|X| \leq |Y|$ a.s. Then $X \in \mathbb{L}^1$ and $\mathbb{E}|X| \leq \mathbb{E}|Y|$. In particular, if X is bounded a.s., then $X \in \mathbb{L}^1$.

Remark 0.10 *It is easy to see that $\mathbb{L}^1$ is a vector space over $\mathbb{R}$ if we define the null vector $X = 0$ if $X = 0$ a.s. In fact $\mathbb{L}^1$ is a Banach space with respect to the norm $\|X\|_1 = \mathbb{E}|X|$, $X \in \mathbb{L}^1$.*

Let X be a random variable on $(\Omega, \mathcal{F}, \mathbb{P})$, and let $g(X)$ be integrable over Ω with respect to $\mathbb{P}$. We next study some particular forms of g which will be used subsequently.

(a) Let $g(x) = x^n$, where n is a positive integer. Then $\alpha_n = \mathbb{E}X^n$, if it exists, is called the moment of order n of the random variable X.

(b) Let $g(x) = |x|^\lambda$, where λ is a positive real number. Then $\beta_\lambda = \mathbb{E}|X|^\lambda$, if it exists, is called the absolute moment of order λ of the random variable X.

(c) Let $g(x) = (x - \gamma)^n$, where γ is a real number and n is a positive integer. Then $\mathbb{E}(X - \gamma)^n$, if it exists, is known as the moment of order n about the point γ. In particular, if $\gamma = \mathbb{E}X$, then $\mathbb{E}(X - \mathbb{E}X)^n$ is called the central moment of order n and is denoted by μ_n. Clearly $\mu_1 = 0$. For $n = 2$,

$$\mu_2 = \mathbb{E}(X - \mathbb{E}X)^2 = \mathbb{E}X^2 - (\mathbb{E}X)^2 \tag{0.4}$$

is called the variance of X and is denoted by $var(X)$. The positive square root of $var(X)$ is called the standard deviation of X. We note that $var(X) \geq 0$, and $var(X) = 0 \Leftrightarrow X = c$ a.s., where c is a constant.

(d) Let $g(x) = e^{tx}$, $t \in (-\delta, \delta)$ for some positive real number δ. Then we write $M(t) = \mathbb{E}e^{tX}$, if it exists, and call it the moment generating function of the random variable X.

(e) Let $g(x) = e^{itx} = \cos tx + i \sin tx$, where $t \in \mathbb{R}$ and $i = \sqrt{-1}$. Then we write

$$\varphi(t) = \mathbb{E}e^{itX} = \mathbb{E} \cos tX + i\mathbb{E} \sin tX,$$

and call φ the characteristic function of the random variable X. We note that the characteristic function φ of a random variable X always exists.

In the following examples we briefly introduce some standard distributions.

Example 0.3 *Let X be a random variable with binomial distribution given by*

$$p(k) = \mathbb{P}\{X = k\} = C_n^k p^k (1 - p)^{n-k}, \quad k = 0, 1, \ldots, n,$$

where $0 < p < 1$. Then moments of all order exist. We have

$$\mathbb{E}X = np,$$

$$var(X) = np(1 - p).$$

Also,

$$\varphi(t) = \mathbb{E}e^{itX}$$
$$= [(1 - p) + pe^{it}]^n.$$

Example 0.4 *Let X be a random variable with Poisson distribution*

$$p(k) = \mathbb{P}\{X = k\} = \frac{e^{-\lambda}\lambda^k}{k!}, \quad k = 0,\ 1,\ 2,\ \ldots,$$

where $\lambda > 0$. In this case

$$\mathbb{E}X = \sum_{k=0}^{\infty} \frac{ke^{-\lambda}\lambda^k}{k!} = \lambda,$$

$$var(X) = \lambda$$

and

$$\varphi(t) = \mathbb{E}e^{itX}$$
$$= \exp\{\lambda(e^{it} - 1)\}.$$

Example 0.5 *If the random variable X has a uniform distribution on (a, b), then*

$$\mathbb{E}X = \frac{a + b}{2},$$

$$var(X) = \frac{(b - a)^2}{12},$$

and, if $t \neq 0$,

$$\varphi(t) = \mathbb{E}e^{itX}$$
$$= \frac{1}{t(b - a)}(e^{itb} - e^{ita}).$$

Example 0.6 *Let the random variable X have the normal distribution given by the probability density function*

$$f(x) = \left(2\pi\sigma^2\right)^{-\frac{1}{2}} \exp\left\{-(x - \mu)^2/2\sigma^2\right\}, \quad x \in \mathbb{R},$$

where $\sigma^2 > 0$ and $\mu \in \mathbb{R}$. Then $\mathbb{E}|X|^\gamma < \infty$ for every $\gamma > 0$, and we have

$$\mathbb{E}X = \mu$$

and

$$var(X) = \sigma^2.$$

Also,

$$\varphi(t) = \mathbb{E}e^{itX}$$
$$= e^{it\mu}e^{-t^2\sigma^2/2}, \quad t \in \mathbb{R}.$$

Example 0.7 *Let the random variable X have a Cauchy probability density function*

$$f(x) = \frac{1}{\pi(1+x^2)}, \quad x \in \mathbb{R}.$$

In this case $\int_{-\infty}^{\infty}|x|f(x)\mathrm{d}x = \infty$, so that $\mathbb{E}X$ does not exist. Also,

$$\varphi(t) = \mathbb{E}e^{itX}$$
$$= \frac{1}{\pi}\int_{-\infty}^{\infty}\frac{e^{itx}}{1+x^2}\mathrm{d}x$$
$$= e^{-|t|}, \quad t \in \mathbb{R}.$$

Example 0.8 *Let X be a random variable with gamma probability density function given by*

$$f(x) = \begin{cases} \dfrac{x^{\alpha-1}e^{-x/\beta}}{\Gamma(\alpha)\beta^{\alpha}} & \text{if } x > 0, \\ 0 & \text{if } x \le 0, \end{cases}$$

where $\alpha > 0$, $\beta > 0$. Then

$$\mathbb{E}X = \alpha\beta$$

and

$$var(X) = \alpha\beta^2.$$

Also,

$$\varphi(t) = \mathbb{E}e^{itX}$$
$$= (1 - i\beta t)^{-\alpha}, \quad t \in \mathbb{R}.$$

We first consider some inequalities concerning moments. The Markov and the Cauchy-Schwartz inequalities, in particular, are among the most commonly used inequalities in probability theory.

Proposition 0.5 *Suppose that $\mathbb{E}|X|^{\lambda} < \infty$ for some $\lambda > 0$. Then $\mathbb{E}|X|^{\nu} < \infty$ for $0 \le \nu \le \lambda$.*

Proposition 0.6 *Let X be a random variable, and let g be a nonnegative Borel-measurable function such that $\mathbb{E}g(X) < \infty$. Suppose that g is even and nondecreasing on $[0,\infty)$. Then, for every $\varepsilon > 0$,*

$$\mathbb{P}\{|X| \ge \varepsilon\} \le \frac{\mathbb{E}g(X)}{g(\varepsilon)}. \tag{0.5}$$

Corollary 0.2 *(Markov's Inequality) If $\mathbb{E}|X|^{\lambda} < \infty$ for some $\lambda > 0$, then, for every $\varepsilon > 0$,*

$$\mathbb{P}\{|X| \ge \varepsilon\} \le \frac{\mathbb{E}|X|^{\lambda}}{\varepsilon^{\lambda}}. \tag{0.6}$$

In particular, if $\lambda = 2$, we get

$$\mathbb{P}\{|X| \geq \varepsilon\} \leq \frac{\mathbb{E}X^2}{\varepsilon^2} \quad (\varepsilon > 0), \tag{0.7}$$

which is known as the Chebyshev inequality.

Proposition 0.7 *Let X and g be as in Proposition 0.6, and suppose further that a.s. $\sup g(x) < \infty$. Then the inequality*

$$\mathbb{P}\{|X| \geq \varepsilon\} \geq \frac{\mathbb{E}g(X) - g(\varepsilon)}{a.s. \ \sup g(X)} \tag{0.8}$$

holds for every $\varepsilon > 0$.

Corollary 0.3 *For $\lambda > 0$ and every $\varepsilon > 0$ the following inequality holds:*

$$\mathbb{E}\frac{|X|^\lambda}{1 + |X|^\lambda} - \frac{\varepsilon^\lambda}{1 + \varepsilon^\lambda} \leq \mathbb{P}\{|X| \geq \varepsilon\} \leq \frac{1 + \varepsilon^\lambda}{\varepsilon^\lambda} \mathbb{E}\frac{|X|^\lambda}{1 + |X|^\lambda}. \tag{0.9}$$

Proposition 0.8 (*Cauchy-Schwartz Inequality*) *Let X and Y be any two random variables with $\mathbb{E}X^2 < \infty$ and $\mathbb{E}Y^2 < \infty$. Then $\mathbb{E}|XY| < \infty$, and the following inequality holds:*

$$(\mathbb{E}|XY|)^2 \leq \mathbb{E}X^2 \cdot \mathbb{E}Y^2. \tag{0.10}$$

Equality (0.10) holds if and only if Y is a linear function of X a.s.

Corollary 0.4 (*Lyapounov's Inequality*) *Let X be a random variable with $\beta_n < \infty$ for some positive integer n. Then*

$$\beta_{k-1}^{1/(k-1)} \leq \beta_k^{1/k}, \quad k = 2, \ 3, \ \ldots, \ n. \tag{0.11}$$

Proposition 0.9 (*Holder's Inequality*) *Let p and q be two positive real numbers satisfying $1 < p, \ q < \infty$, and $p^{-1} + q^{-1} = 1$. Let X and Y be two random variables such that $\mathbb{E}|X|^p < \infty$ and $\mathbb{E}|Y|^q < \infty$. Then $\mathbb{E}|XY| < \infty$, and the inequality*

$$\mathbb{E}|XY| \leq (\mathbb{E}|X|^p)^{1/p} (\mathbb{E}|Y|^q)^{1/q} \tag{0.12}$$

holds.

Remark 0.11 *Proposition 0.8 is a special case of Proposition 0.9 where $p = q = 2$.*

Proposition 0.10 (*Minkowski's Inequality*) *Let $1 \leq p < \infty$. Let X and Y be two random variables such that $\mathbb{E}|X|^p < \infty$ and $\mathbb{E}|Y|^p < \infty$. Then $\mathbb{E}|X + Y|^p < \infty$, and the inequality*

$$(\mathbb{E}|X + Y|^p)^{1/p} \leq (\mathbb{E}|X|^p)^{1/p} + (\mathbb{E}|Y|^p)^{1/p} \tag{0.13}$$

holds.

Remark 0.12 *Let $1 \leq p < \infty$, and $\mathbb{L}^p$ be the set of all random variables X on Ω with $\mathbb{E}|X|^p < \infty$. Then it follows immediately from Proposition 0.10 that the set $\mathbb{L}^p$ is a vector space over $\mathbb{R}$ if we define the null vector $X = 0$ if $X = 0$ a.s. In fact $\mathbb{L}^p$ is a Banach space with respect to the norm $\|X\|_p = (\mathbb{E}|X|^p)^{1/p}$, $X \in \mathbb{L}^p$.*

We next give some condition for the existence of the moments of a random variable in terms of its distribution function.

Proposition 0.11 *Let X be a random variable defined on $(\Omega, \mathcal{F}, \mathbb{P})$. Then*

$$\mathbb{E}|X|^p < \infty \Rightarrow \lim_{x \to +\infty} x^p \mathbb{P}\{|X| \geq x\} = 0.$$

Proposition 0.12 *Let the random variable $X \geq 0$ a.s., and let F be its distribution function. Then*

$$\mathbb{E}X < \infty \Leftrightarrow \int_0^{+\infty} [1 - F(x)]\mathrm{d}x < \infty.$$

In this case the relation

$$\mathbb{E}X = \int_0^{+\infty} [1 - F(x)]\mathrm{d}x. \tag{0.14}$$

holds.

Corollary 0.5 *Let X be a random variable with distribution function F. Then $\mathbb{E}|X| < \infty$ if and only if both the integrals $\int_{-\infty}^0 F(x)\mathrm{d}x$ and $\int_0^{+\infty}[1 - F(x)]\mathrm{d}x$ are finite. In this case*

$$\mathbb{E}X = \int_0^{+\infty} [1 - F(x)]\mathrm{d}x - \int_{-\infty}^0 F(x)\mathrm{d}x. \tag{0.15}$$

Corollary 0.6 (*Moments Lemma*) *Let X be a random variable, and let $0 < p < \infty$. Then*

$$\mathbb{E}|X|^p < \infty \Leftrightarrow \sum_{n=1}^{\infty} \mathbb{P}\{|X| \geq n^{1/p}\} < \infty.$$

Corollary 0.7 *Let X be a random variable satisfying $\lim_{n \to \infty} n^p \mathbb{P}\{|X| \geq n\} = 0$ for some $p > 0$. Then $\mathbb{E}|X|^q < \infty$ for $0 \leq q < p$.*

Let $\mathbb{B}$ be a real Banach space with norm $\|\cdot\|$. The expected value of a $\mathbb{B}$-valued random element, where $\mathbb{B}$ is separable, is defined in terms of a Pettis integral.

Definition 0.7 *Let $\mathbb{B}$ be a separable Banach space, and let X be a $\mathbb{B}$-valued random element. We say that the expected value of X exists if the following conditions*

are met:
(a) $\mathbb{E}|l(X)| < \infty$ *for all* $l \in \mathbb{B}^*$.
(b) *There exists an element* $\mathbb{E}X \in \mathbb{B}$ *such that the relation*

$$l(\mathbb{E}X) = \mathbb{E}l(X) = \int_\Omega l(X)\mathrm{d}\mathbb{P} \qquad (0.16)$$

holds for all $l \in \mathbb{B}^*$. *Here* $\mathbb{E}X$ *is called the expected value of* X *and is unique.*

We note that $\mathbb{E}X$ exists if $\mathbb{E}\|X\| < \infty$. The following proposition gives some properties of $\mathbb{E}X$.

Proposition 0.13 *Let* $\mathbb{B}$ *be a separable Banach space, and let* X *be a* $\mathbb{B}$-*valued random element such that* $\mathbb{E}X$ *exists. Then the following conditions holds:*
(a) *For every* $\alpha \in \mathbb{R}$, $\mathbb{E}(\alpha X) = \alpha\mathbb{E}X$.
(b) *If* $\mathbb{P}\{X = x\} = 1$, *then* $\mathbb{E}X = x$; *moreover, if in addition* ζ *is a real-valued random variable such that* $\mathbb{E}|\zeta| < \infty$, *then* $\mathbb{E}(\zeta X) = x\mathbb{E}\zeta$.
(c) *Let* T *be a bounded (continuous) linear operator from* $\mathbb{B}$ *to* $\mathbb{B}_0$, *where* $\mathbb{B}_0$ *is a Banach space. Then* $\mathbb{E}T(X)$ *exists, and the relation*

$$\mathbb{E}T(X) = T(\mathbb{E}X)$$

holds.
(d) *The relation*

$$\|\mathbb{E}X\| \leq \mathbb{E}\|X\|$$

holds where $\mathbb{E}\|X\|$ *may be infinite.*
(e) *Let* Y *be a* $\mathbb{B}$-*valued random element such that* $\mathbb{E}Y$ *exists. Then* $\mathbb{E}(X+Y)$ *also exists, and the relation* $\mathbb{E}(X+Y) = \mathbb{E}X + \mathbb{E}Y$ *holds.*

Example 0.9 *Let* $\mathbb{B} = c$ *be the set of all convergent sequences* $x = (x_1, \ x_2, \ \ldots)$ *of real numbers with norm* $\|x\| = \sup_{n\geq 1}|x_n|$. *Then* $\mathbb{B}$ *is a Banach space. Let* X *be a* $\mathbb{B}$-*valued random element defined on some probability space* $(\Omega, \mathcal{F}, \mathbb{P})$. *For* $\omega \in \Omega$ *let* $X(\omega) = (X_1(\omega), \ X_2(\omega), \ \ldots)$. *Then* $\{X_n(\omega)\}$ *is a convergent sequence of real numbers for each* $\omega \in \Omega$. *Suppose that*

$$\mathbb{E}\|X\| = \mathbb{E}\sup_{n\geq 1}|X_n| < \infty.$$

Then $\mathbb{E}X = (\mathbb{E}X_1, \ \mathbb{E}X_2, \ \ldots)$. *In fact, note that* $\lim_{n\to\infty} X_n$ *exists a.s. and* $\sup_{n\geq 1} X_n$ *has a finite expectation. It follows from the Lebesgue dominated convergence theorem that* $\lim_{n\to\infty}\mathbb{E}X_n$ *exists. Hence* $(\mathbb{E}X_1, \ \mathbb{E}X_2, \ \ldots) \in \mathbb{B}$. *Next note that the dual* $\mathbb{B}^*$ *consists of all summable sequences* $l = (l_0, \ l_1, \ l_2, \ \ldots)$,

$\sum_{n=0}^{\infty} |l_n| < \infty$. *Hence for any $l \in \mathbb{B}^*$ we have*

$$l(\mathbb{E}X) = l_0 \lim_{n\to\infty} \mathbb{E}X_n + \sum_{n=1}^{\infty} l_n \mathbb{E}X_n$$

$$= \mathbb{E}\left\{ l_0 \lim_{n\to\infty} X_n + \sum_{n=1}^{\infty} l_n X_n \right\}$$

$$= \mathbb{E}l(X)$$

so that (0.16) holds. It follows that $\mathbb{E}X = (\mathbb{E}X_1,\ \mathbb{E}X_2,\ \ldots)$.

Example 0.10 *Let $\mathbb{B} = C[0,1]$, the set of all continuous real-valued functions on the interval $[0,1]$ with norm*

$$\|x\| = \sup_{0 \le t \le 1} |x(t)|,$$

where $x = \{x(t),\ 0 \le t \le 1\} \in \mathbb{B}$. Then $\mathbb{B}$ is a separable Banach space. Let $X = \{X_t,\ 0 \le t \le 1\}$ be a $\mathbb{B}$-valued random element. Suppose that

$$\mathbb{E}\|X\| = \mathbb{E} \sup_{0 \le t \le 1} |X_t| < \infty.$$

We show that $\mathbb{E}X = \{\mathbb{E}X_t,\ 0 \le t \le 1\}$. First note that

$$\lim_{h\to 0} |\mathbb{E}X_{t+h} - \mathbb{E}X_t| \le \lim_{h\to 0} \mathbb{E}|X_{t+h} - X_t| = 0$$

in view of the Lebesgue dominated convergence theorem. It follows that $\{\mathbb{E}X_t,\ 0 \le t \le 1\} \in \mathbb{B}$. Now $\mathbb{B}^*$ consists of finite signed measures μ on $[0,1]$, so for any $\mu \in \mathbb{B}^*$ we have

$$\mu(\mathbb{E}X) = \int_0^1 \mathbb{E}X_t d\mu(t) = \mathbb{E} \int_0^1 X_t d\mu(t) = \mathbb{E}\mu(X),$$

and it follows that $\mathbb{E}X = \{\mathbb{E}X_t,\ 0 \le t \le 1\}$.

Definition 0.8 *Let $\mathbb{B}$ be a separable Banach space, and let X be a $\mathbb{B}$-valued random element. We say that the variance of X exists if $\mathbb{E}X$ exists and*

$$var(X) = \int_\Omega \|X - \mathbb{E}X\|^2 d\mathbb{P} < \infty.$$

Here $var(X)$ is known as the variance of X, and its nonnegative square root $\sigma(X)$ is called the standard deviation of X.

Remark 0.13 *Let X be a $\mathbb{B}$-valued random element such that $var(X)$ exists. Then the following inequality (Chebyshev's inequality):*

$$\mathbb{P}\{\|X - \mathbb{E}X\| \geq \varepsilon\} < \varepsilon^{-2} var(X) \tag{0.17}$$

holds for every $\varepsilon > 0$. The proof of (0.17) is an immediate consequence of Proposition 0.4, the Chebyshev inequality for real-valued random variables, and the definition of $var(X)$.

Remark 0.14 *Moments of any order p, $1 \leq p < \infty$, can be defined in a similar manner.*

In the following we study the abstract notions of conditional probability and conditional expectation, which play a key role in probability theorey (including the theory of stochastic processes) and mathematical statistics.

Let $(\Omega, \mathcal{F}, \mathbb{P})$ be a probability space. Let $A \in \mathcal{F}$ and $B \in \mathcal{F}$. Then the conditional pobability of A, given B, is defined to be

$$\mathbb{P}\{A|B\} = \frac{\mathbb{P}\{A \cap B\}}{\mathbb{P}\{B\}} \tag{0.18}$$

provided that $\mathbb{P}\{B\} > 0$. Clearly

$$\mathbb{P}\{A \cap B\} = \mathbb{P}\{B\}\mathbb{P}\{A|B\}.$$

In a similar manner the conditional probability of B, given A, namely, $\mathbb{P}\{B|A\}$, can be defined, provided that $\mathbb{P}\{A\} > 0$. We note that $A \in \mathcal{F}$ and $B \in \mathcal{F}$, with $\mathbb{P}\{A\} > 0$, $\mathbb{P}\{B\} > 0$, are independent if and only if either one of the following two equivalent conditions holds:

$$\mathbb{P}\{A|B\} = \mathbb{P}\{A\},$$

$$\mathbb{P}\{B|A\} = \mathbb{P}\{B\}.$$

Let $B \in \mathcal{F}$ with $\mathbb{P}\{B\} > 0$ be fixed. Then $\mathbb{P}\{\cdot|B\}$ defined in (0.18) is a probability measure on $(\Omega, \mathcal{F})$, so that $(\Omega, \mathcal{F}, \mathbb{P}\{\cdot|B\})$ is a probability space.

Let X be a random variable defined on $(\Omega, \mathcal{F}, \mathbb{P})$ such that $\mathbb{E}X$ exists. Clearly X is also integrable with respect to $\mathbb{P}\{\cdot|B\}$ on Ω. We set

$$\mathbb{E}[X|B] = \int_{\Omega} X d\mathbb{P}\{\cdot|B\}, \tag{0.19}$$

and in this case $\mathbb{E}[X|B]$ is known as the conditional expectation of X, given B. Since

$$\mathbb{P}\{\cdot|B\} = 0 \quad \text{on the class } \{A \cap B^c : A \in \mathcal{F}\}$$

and

$$\mathbb{P}\{\cdot|B\} = \frac{\mathbb{P}\{\cdot\}}{\mathbb{P}\{B\}} \quad \text{on the class } \{A \cap B : A \in \mathcal{F}\},$$

we can rewrite (0.19) as

$$\mathbb{E}[X|B] = \frac{1}{\mathbb{P}\{B\}} \int_B X d\mathbb{P} = \frac{1}{\mathbb{P}\{B\}} \mathbb{E}[X\mathbf{1}_B], \tag{0.20}$$

where $\mathbf{1}_B$ is the indicator function of event B. In particular, if $X = \mathbf{1}_A$ for $A \in \mathcal{F}$, then

$$\mathbb{E}[\mathbf{1}_A|B] = \frac{\mathbb{P}\{A \cap B\}}{\mathbb{P}\{B\}} = \mathbb{P}\{A|B\}.$$

Consider the σ-field $\mathcal{U} = \{\Omega, B, B^c, \emptyset\} \subset \mathcal{F}$ generated by $B \in \mathcal{F}$. By setting

$$\mathbb{E}[X|\mathcal{U}] = \begin{cases} \mathbb{E}[X|B] & \text{if } \omega \in B, \\ \mathbb{E}[X|B^c] & \text{if } \omega \in B^c, \end{cases}$$

we see that $\mathbb{E}[X|\mathcal{U}]$ is a two-point random variable which satisfies

$$\begin{aligned} \int_\Omega \mathbb{E}[X|\mathcal{U}] d\mathbb{P} &= \mathbb{P}\{B\}\mathbb{E}[X|B] + \mathbb{P}\{B^c\}\mathbb{E}[X|B^c] \\ &= \mathbb{E}[X\mathbf{1}_B] + \mathbb{E}[X\mathbf{1}_{B^c}] \\ &= \int_\Omega X d\mathbb{P}. \end{aligned}$$

The $\mathcal{U}$-measurable function $\mathbb{E}[X|\mathcal{U}]$ defined on Ω is known as the conditional expectation of X, given (the σ-field) $\mathcal{U}$.

We next consider the σ-field $\mathcal{U} \subset \mathcal{F}$, generated by a finite number of sets $B_1, B_2, \ldots, B_n \in \mathcal{F}$ ($n \geq 2$) such that $B_i \cap B_j = \emptyset$ for $i \neq j$, $\bigcup_{i=1}^n B_i = \Omega$ and $\mathbb{P}\{B_i\} > 0$ for $i = 1, 2, \ldots, n$. Let X be a random variable such that $\mathbb{E}X$ exists. Then we set

$$\mathbb{E}[X|\mathcal{U}] = \mathbb{E}[X|B_i] \quad \text{if } \omega \in B_i, \quad i = 1, 2, \ldots, n, \tag{0.21}$$

where $\mathbb{E}[X|B_i]$ is as defined in (0.19). Then $\mathbb{E}[X|\mathcal{U}]$ is called the conditional expectation of X, given (the σ-field) $\mathcal{U}$. Clearly $\mathbb{E}[X|\mathcal{U}]$ is a random variable. Using (0.20), we see that

$$\mathbb{P}\{B_i\}\mathbb{E}[X|B_i] = \mathbb{E}[X\mathbf{1}_{B_i}], \quad i = 1, 2, \ldots, n,$$

so that

$$\begin{aligned} \sum_{i=1}^n \mathbb{P}\{B_i\}\mathbb{E}[X|B_i] &= \sum_{i=1}^n \mathbb{E}[X\mathbf{1}_{B_i}] \\ &= \mathbb{E}X \\ &= \int_\Omega X d\mathbb{P}. \end{aligned}$$

On the other hand, $\mathbb{E}[X|\mathcal{U}]$ is a simple random variable taking values $\mathbb{E}[X|B_i]$ on B_i for $i = 1, 2, \ldots, n$, so that it follows from the definition of the integral of a simple function that

$$\int_\Omega \mathbb{E}[X|\mathcal{U}]\mathrm{d}\mathbb{P} = \sum_{i=1}^n \mathbb{P}\{B_i\}\mathbb{E}[X|B_i] = \int_\Omega X\mathrm{d}\mathbb{P}. \qquad (0.22)$$

Clearly $\mathbb{E}[X|\mathcal{U}]$ is measurable with respect to $\mathcal{U}$. We see easily that, if X itself is $\mathcal{U}$-measurable, $\mathbb{E}[X|\mathcal{U}] = X$ a.s. with respect to the probability measure induced by $\mathbb{P}$ on $\mathcal{U}$.

More generally, let $\{B_n : n \geq 1\} \subset \mathcal{F}$ be a countable partition of Ω with $\mathbb{P}\{B_n\} > 0$ for $n \geq 1$, and let $\mathcal{U}$ be the σ-field generated by the B_n. Let X be a random variable with finite expectation. Then

$$\mathbb{E}[X|\mathcal{U}] = \sum_{n=1}^\infty \mathbb{E}[X|B_n]\mathbf{1}_{B_n} \qquad (0.23)$$

defines the conditional expectation of X, given $\mathcal{U}$. As before, $\mathbb{E}[X|\mathcal{U}]$ is a discrete random variable taking a countable set of values $\{\mathbb{E}[X|B_n], \; n \geq 1\}$, and it follows that

$$\int_\Omega \mathbb{E}[X|\mathcal{U}]\mathrm{d}\mathbb{P} = \int_\Omega X\mathrm{d}\mathbb{P}. \qquad (0.24)$$

Clearly $\mathbb{E}[X|\mathcal{U}]$ is measurable with respect to $\mathcal{U}$; in particular, if X itself is $\mathcal{U}$-measurable, $\mathbb{E}[X|\mathcal{U}] = X$ a.s. with respect to the probability measure induced by $\mathbb{P}$ on $\mathcal{U}$.

In practical applications we need the concept of conditional expectation of a random variable X, given a σ-field generated by a random variable Y or, more generally, by a fixed collection of random variables $\{Y_\lambda, \; \lambda \in \Lambda\}$. For this purpose it is necessary to extend the definition of conditional expectation of a random variable to the case where the given σ-field is arbitrary.

Let $(\Omega, \mathcal{F}, \mathbb{P})$ be a probability space, and let $\mathcal{D} \subset \mathcal{F}$ be a σ-field. Let $\mathbb{P}_\mathcal{D}$ be the probability measure induced by $\mathbb{P}$ on $\mathcal{D}$, that is, $\mathbb{P}_\mathcal{D}\{E\} = \mathbb{P}\{E\}$ for $E \in \mathcal{D}$. Let X be a random variable defined on $(\Omega, \mathcal{F}, \mathbb{P})$ such that $\mathbb{E}X$ exists. Then for every $E \in \mathcal{D}$ we can define the indefinite integral

$$Q_X\{E\} = \int_E X\mathrm{d}\mathbb{P} = \int_\Omega X\mathbf{1}_E\mathrm{d}\mathbb{P}.$$

Clearly Q_X is a finite signed measure on $\mathcal{D}$ such that $Q_X\{E\} = 0$ for every $E \in \mathcal{D}$ for which $\mathbb{P}_\mathcal{D}\{E\} = 0$. Hence Q_X is absolutely continuous with respect to $\mathbb{P}_\mathcal{D}$, so that in view of the Radon-Nikodym theorem there exists a $\mathcal{D}$-measurable function defined on Ω, which we denote by $\mathbb{E}[X|\mathcal{D}]$, such that the relation

$$\int_E \mathbb{E}[X|\mathcal{D}]\mathrm{d}\mathbb{P}_\mathcal{D} = Q_X\{E\} = \int_E X\mathrm{d}\mathbb{P} \qquad (0.25)$$

holds for every $E \in \mathcal{D}$. Here the function $\mathbb{E}[X|\mathcal{D}]$ is determined uniquely with respect to $\mathbb{P}_{\mathcal{D}}$ in the sense that, if there exists another $\mathcal{D}$-measurable function g on Ω satisfying (0.25) for every $E \in \mathcal{D}$, then $g = \mathbb{E}[X|\mathcal{D}]$ a.s. with respect to $\mathbb{P}_{\mathcal{D}}$.

Definition 0.9 *The $\mathcal{D}$-measurable function $\mathbb{E}[X|\mathcal{D}]$ defined by (0.25) is said to be the conditional expectation of the random variable X with respect to the σ-field $\mathcal{D}$. Here $\mathbb{E}[X|\mathcal{D}]$ is defined uniquely except for $\mathcal{D}$-measurable sets of $\mathbb{P}_{\mathcal{D}}$-measure zero.*

We note that

$$\int_{\Omega} \mathbb{E}[X|\mathcal{D}] d\mathbb{P}_{\mathcal{D}} = \int_{\Omega} X d\mathbb{P} = \mathbb{E}X \qquad (0.26)$$

In particular, let Y be another random variable defined on Ω, and let $\mathcal{D} = \sigma(Y)$, where $\sigma(Y)$ is the σ-field generated by Y. Then we write $\mathbb{E}[X|Y] = \mathbb{E}[X|\mathcal{D}]$. More generally, let $\{Y_\lambda, \ \lambda \in \Lambda\}$ be a collection of random variables defined on Ω, and let $\mathcal{D} = \sigma(\{Y_\lambda, \ \lambda \in \Lambda\})$. Then we write $\mathbb{E}[X|Y_\lambda, \ \lambda \in \Lambda] = \mathbb{E}[X|\mathcal{D}]$.

The $\mathcal{D}$-measurable function $\mathbb{E}[X|\mathcal{D}]$ is determined uniquely up to $\mathcal{D}$-measurable null set by condition (0.25). The class of all $\mathcal{D}$-measurable function satisfying (0.25) for all $D \in \mathcal{D}$ is an equivalence class under the relation of a.s. equality. Any member of this class is known as a version of the conditional expectation of X, given (or with respect to) $\mathcal{D}$. Note that, although $\mathbb{E}[X|\mathcal{D}]$ is a.s. uniquely determined on Ω, the exceptional set may depend on the random variable X under consideration.

Example 0.11 *If $\mathcal{D} = \{\emptyset, \Omega\}$, only constant functions are $\mathcal{D}$-measurable, and it follows that $\mathbb{E}[X|\mathcal{D}] = \mathbb{E}X$ for all $\omega \in \Omega$. If $\mathcal{D} = \mathcal{F}$, then X is $\mathcal{D}$-measurable and $\mathbb{E}[X|\mathcal{D}] = X$ a.s.*

Example 0.12 *Let $\Omega = (0,1)$, $\mathcal{B} = \mathcal{B}_{(0,1)}$ be the Borel σ-field on Ω, and λ be the Lebesgue measure on $\mathcal{B}$. Let*

$$Y(\omega) = \begin{cases} 2 & \text{if } 0 < \omega < \dfrac{1}{3}, \\ 5 & \text{if } \dfrac{1}{3} \leq \omega < 1. \end{cases}$$

Let $\mathcal{D} = \sigma(Y)$. Then $\mathcal{D} = \{\Omega, \emptyset, (0, \frac{1}{3}), [\frac{1}{3}, 1)\}$. Only simple random variables with a (single) jump at $\frac{1}{3}$ are $\mathcal{D}$-measurable. Let $X(\omega) = \omega$ for all $\omega \in (0,1)$. Then

$$\mathbb{E}[X|\mathcal{D}] = \mathbb{E}[X|Y] = \begin{cases} \dfrac{1}{6} & \text{if } \omega \in \left(0, \dfrac{1}{3}\right), \\ \dfrac{2}{3} & \text{if } \omega \in \left[\dfrac{1}{3}, 1\right). \end{cases}$$

We now study some properties of the function $\mathbb{E}[X|\mathcal{D}]$ defined above.

Proposition 0.14 *Let X be a random variable defined on $(\Omega, \mathcal{F}, \mathbb{P})$ such that $\mathbb{E}X$ exists, and let $\mathcal{D} \subset \mathcal{F}$ be a σ-field.*
(a) Let X be $\mathcal{D}$-measurable. Then $\mathbb{E}[X|\mathcal{D}] = X$ a.s. $(\mathbb{P}_\mathcal{D})$.
(b) Let $X = c$ a.s. $(\mathbb{P})$, where c is a constant. Then $\mathbb{E}[X|\mathcal{D}] = c$ a.s. $(\mathbb{P}_\mathcal{D})$.
(c) Let $X \geq 0$ a.s. $(\mathbb{P})$. Then $\mathbb{E}[X|\mathcal{D}] \geq 0$ a.s. $(\mathbb{P}_\mathcal{D})$.
(d) Let Y be another random variable on $(\Omega, \mathcal{F}, \mathbb{P})$ such that $\mathbb{E}Y$ exists. Let $a,\ b \in \mathbb{R}$. Then

$$\mathbb{E}[aX + bY|\mathcal{D}] = a\mathbb{E}[X|\mathcal{D}] + b\mathbb{E}[Y|\mathcal{D}] \quad a.s. \ (\mathbb{P}_\mathcal{D}).$$

Proposition 0.15 *(Conditional Monotone Convergence Theorem) Let X_n be a nondecreasing sequence of nonnegative random variables which converge a.s. on Ω to a random variable X. Suppose that $\mathbb{E}X$ exists. Then*

$$0 \leq \lim_{n \to \infty} \mathbb{E}[X_n|\mathcal{D}] = \mathbb{E}[X|\mathcal{D}] \quad a.s. \ (\mathbb{P}_\mathcal{D}).$$

Proposition 0.16 *(Conditional Lebesgue Dominated Convergence Theorem) Let X_n be a sequence of random variables which converge a.s. to a random variable X on Ω. Suppose that there exists a random variable $Y \geq 0$ a.s. $(\mathbb{P})$ such that $\mathbb{E}Y$ exists and $|X_n| \leq Y$ a.s. $(\mathbb{P})$ for all $n \geq 1$. Then*

$$\lim_{n \to \infty} \mathbb{E}[X_n|\mathcal{D}] = \mathbb{E}[X|\mathcal{D}] \quad a.s. \ (\mathbb{P}_\mathcal{D}).$$

Proposition 0.17 *(Conditional Fatou's Lemma) Let X_n be a sequence of nonnegative random variables such that $\mathbb{E}X_n$ exists for $n \geq 1$ and $\liminf_{n \to \infty} \mathbb{E}X_n < \infty$. Then*

$$\mathbb{E}\left[\liminf_{n \to \infty} X_n \Big| \mathcal{D}\right] \leq \liminf_{n \to \infty} \mathbb{E}[X_n|\mathcal{D}] \quad a.s. \ (\mathbb{P}_\mathcal{D}).$$

The following result is quite useful.

Proposition 0.18 *Let X and Y be two random variables such that $\mathbb{E}[XY]$ and $\mathbb{E}Y$ exist. Suppose that X be $\mathcal{D}$-measurable. Then*

$$\mathbb{E}[XY|\mathcal{D}] = X\mathbb{E}[Y|\mathcal{D}] \quad a.s. \ (\mathbb{P}_\mathcal{D}).$$

Corollary 0.8 *Let $\mathcal{D}_1$, $\mathcal{D}_2$ be two sub-σ-fields of $\mathcal{F}$ such that $\mathcal{D}_1 \subset \mathcal{D}_2$. Let X and Y be random variables such that $\mathbb{E}[XY]$ and $\mathbb{E}Y$ exist and X is $\mathcal{D}_2$-measurable. Then*

$$\mathbb{E}[XY|\mathcal{D}_1] = \mathbb{E}\big[X\mathbb{E}[Y|\mathcal{D}_2]\big|\mathcal{D}_1\big] \quad a.s. \ (\mathbb{P}_{\mathcal{D}_1}).$$

Corollary 0.9 *Let $\mathcal{D}_1 \subset \mathcal{D}_2 \subset \mathcal{F}$ be two sub-σ-fields. If $\mathbb{E}Y$ exists, then*

$$\mathbb{E}[Y|\mathcal{D}_1] = \mathbb{E}\big[\mathbb{E}[Y|\mathcal{D}_1]\big|\mathcal{D}_2\big] = \mathbb{E}\big[\mathbb{E}[Y|\mathcal{D}_2]\big|\mathcal{D}_1\big] \quad a.s. \ (\mathbb{P}_{\mathcal{D}_1}).$$

Proposition 0.19 *Let $\mathcal{D} \subset \mathcal{F}$ be a σ-field, and let X be a random variable such that $\mathbb{E}X$ exists and $\sigma(X)$ and $\mathcal{D}$ are independent. Then*

$$\mathbb{E}[X|\mathcal{D}] = \mathbb{E}X \quad a.s. \ (\mathbb{P}_\mathcal{D}).$$

From Proposition 0.19 it follows that, if X and Y are independent, $\mathbb{E}[X|Y] = \mathbb{E}X$ a.s. Let X and Z be independent random variables, and let X be integrable. Does it necessarily follow that

$$\mathbb{E}[X|Y, \; Z] = \mathbb{E}[X|Y] \quad \text{a.s. } (\mathbb{P}_D)?$$

That the answer is no in general (unless both X and Y are independent of Z) is demonstrated in the following example.

Example 0.13 *Let* $\Omega = [0, 1]$, $\mathcal{F} = \mathcal{B}_\Omega$, *the Borel σ-field of subsets of Ω, and $\mathbb{P}$ be the Lebesgue measure on $[0, 1]$. Let*

$$X(\omega) = \begin{cases} 1 & \text{if } \omega \in \left[0, \frac{1}{2}\right], \\ 0 & \text{if } \omega \in \left(\frac{1}{2}, 1\right]; \end{cases}$$

$$Y(\omega) = \begin{cases} 1 & \text{if } \omega \in \left[0, \frac{3}{4}\right), \\ 0 & \text{if } \omega \in \left[\frac{3}{4}, 1\right]; \end{cases}$$

and

$$Z(\omega) = \begin{cases} 1 & \text{if } \omega \in \left[\frac{1}{4}, \frac{3}{4}\right], \\ 0 & \text{if } \omega \in \left[0, \frac{1}{4}\right) \cup \left(\frac{3}{4}, 1\right]. \end{cases}$$

Then X and Z are independent. However,

$$\mathbb{E}[X|Y] = \begin{cases} \dfrac{2}{3} & \text{if } \omega \in \left[0, \frac{3}{4}\right), \\ 0 & \text{otherwise,} \end{cases}$$

and

$$\mathbb{E}[X|Y, \; Z] = \begin{cases} 0 & \text{if } \omega \in \left[\frac{3}{4}, 1\right], \\ \dfrac{1}{2} & \text{if } \omega \in \left[\frac{1}{4}, \frac{3}{4}\right), \\ 1 & \text{if } \omega \in \left[0, \frac{1}{4}\right), \end{cases}$$

so that $\mathbb{E}[X|Y] \neq \mathbb{E}[X|Y, \; Z]$.

Proposition 0.20 *Let X be integrable, and $\mathcal{D}_1$, $\mathcal{D}_2$ be two sub-σ-fields of $\mathcal{F}$ such that $\sigma(X)$ and $\mathcal{D}_1$ are independent of $\mathcal{D}_2$. Then*

$$\mathbb{E}[X|\mathcal{D}_1 \vee \mathcal{D}_2] = \mathbb{E}[X|\mathcal{D}_1] \quad \text{a.s. } (\mathbb{P}_{\mathcal{D}_1}),$$

where $\mathcal{D}_1 \vee \mathcal{D}_2$ is the σ-field generated by $\mathcal{D}_1 \cup \mathcal{D}_2$.

Proposition 0.21 (*Jensen's Inequality*) *Let X be a random variable, defined on $(\Omega, \mathcal{F}, \mathbb{P})$, such that $\mathbb{E}X$ exists. Let g be a (continuous) convex function on $\mathbb{R}$ such that $\mathbb{E}g(X)$ exists. Let $\mathcal{D} \subset \mathcal{F}$ be a σ-field. Then*

$$g\big(\mathbb{E}[X|\mathcal{D}]\big) \leq \mathbb{E}[g(X)|\mathcal{D}] \quad a.s. \ (\mathbb{P}_{\mathcal{D}}).$$

Corollary 0.10 *If g is convex and $\mathbb{E}X$ and $\mathbb{E}g(X)$ both exist, then*

$$g(\mathbb{E}X) \leq \mathbb{E}g(X).$$

Corollary 0.11 *Let $p \geq 1$, and suppose that $\mathbb{E}|X|^p < \infty$. Then*

$$|\mathbb{E}[X|\mathcal{D}]|^p \leq \mathbb{E}[|X|^p|\mathcal{D}] \quad a.s. \ (\mathbb{P}_{\mathcal{D}}).$$

Corollary 0.12 *If $\mathbb{E}X$ exists, then*

$$\big[\mathbb{E}[X|\mathcal{D}]\big]_+ \leq \mathbb{E}[X_+|\mathcal{D}] \quad a.s. \ (\mathbb{P}_{\mathcal{D}})$$

and

$$\big[\mathbb{E}[X|\mathcal{D}]\big]_- \leq \mathbb{E}[X_-|\mathcal{D}] \quad a.s. \ (\mathbb{P}_{\mathcal{D}}).$$

Let $(\Omega, \mathcal{F}, \mathbb{P})$ be a probability space, and let $\mathcal{D} \in \mathcal{F}$ be a sub-σ-field. Let $A \in \mathcal{F}$. Then the indicator function $\mathbf{1}_A$ is an $\mathcal{F}$-measurable simple random variable and $\mathbb{E}[\mathbf{1}_A|\mathcal{D}]$ is well defined.

Definition 0.10 *The conditional probability $\mathbb{P}\{A|\mathcal{D}\}$ of an event A, given $\mathcal{D}$, is defined by*

$$\mathbb{P}\{A|\mathcal{D}\} = \mathbb{E}[\mathbf{1}_A|\mathcal{D}].$$

Clearly, for every $E \in \mathcal{D}$,

$$\int_E \mathbb{P}\{A|\mathcal{D}\}d\mathbb{P}_{\mathcal{D}} = \int_E \mathbf{1}_A d\mathbb{P} = \mathbb{P}\{A \cap E\}. \tag{0.27}$$

We emphasize again that $\mathbb{P}\{A|\mathcal{D}\}$ is $\mathcal{D}$-measurable function which is determined uniquely only up to $\mathcal{D}$-measurable null sets. The class of all $\mathcal{D}$-measurable functions satisfying (0.27) is an equivalence class under the relation of a.s. equality, and any member of this class is known as a version of the conditional probability of A, given $\mathcal{D}$.

It is easy to check that:
(a) $0 \leq \mathbb{P}\{A|\mathcal{D}\} \leq 1$ a.s. and $\mathbb{P}\{\Omega|\mathcal{D}\} = 1$ a.s.
(b) If $A_1, A_2, \ldots$ are disjoint sets in $\mathcal{F}$, then

$$\mathbb{P}\Big\{\bigcup_{n=1}^{\infty} A_n \Big| \mathcal{D}\Big\} = \sum_{n=1}^{\infty} \mathbb{P}\{A_n|\mathcal{D}\} \quad a.s.$$

The results in (a) and (b) do not imply, however, that the set function $A \to \mathbb{P}\{A|\mathcal{D}\}$ for $A \in \mathcal{F}$ is a probability measure on $\mathcal{F}$. The difficulty is that the exceptional (null) set depends on A, so that in (b) the exceptional null set depends on the sequence $\{A_n\}$. The union of these generally uncountably many null sets is usually not a null set. It need not even be in $\mathcal{F}$.

0.3 Convergence and weak convergence

Let $(\Omega, \mathcal{F}, \mathbb{P})$ be a probability space, and let $\{X_n, n \geq 1\}$ be a sequence of random variables defined on it. In this section we consider some useful concepts of convergence of the sequence $\{X_n\}$.

The concept of a.s. convergence in probability theory is identical with the concept of almost everywhere (a.e.) convergence in measure theory.

Definition 0.11 *The sequence of random variables $\{X_n\}$ is said to converge a.s. to the random variables X if and only if there exists a set $E \in \mathcal{F}$ with $\mathbb{P}(E) = 0$ such that, for every $\omega \in E^c$, $|X_n(\omega) - X(\omega)| \to 0$ as $n \to \infty$. In that case we write $X_n \overset{a.s.}{\to} X$.*

Definition 0.12 *The sequence of random variables $\{X_n\}$ is said to Cauchy (fundamental) a.s. if there exists a set $E \in \mathcal{F}$ with $\mathbb{P}(E) = 0$ such that, for every $\omega \in E^c$, $|X_n(\omega) - X_m(\omega)| \to 0$ as $m, n \to \infty$.*

In fact, the two concepts defined above are equivalent.

Proposition 0.22 *Let $\{X_n\}$ be a sequence of random variables defined on $(\Omega, \mathcal{F}, \mathbb{P})$. Then*

$$X_n \overset{a.s.}{\to} X \Leftrightarrow \{X_n\} \text{ is Cauchy a.s.}$$

We note that, if $X_n \overset{a.s.}{\to} X$, then X is also a random variable which is unique a.s. We next give the following criterion for a.s. convergence.

Proposition 0.23 *(a) The sequence of random variables $\{X_n\}$ converges to a random variable X a.s. if and only if*

$$\lim_{n \to \infty} \mathbb{P}\left\{ \bigcup_{m=n}^{\infty} (|X_m - X| \geq \varepsilon) \right\} = 0 \quad \text{for every } \varepsilon > 0.$$

(b) The sequence of random variables $\{X_n\}$ is Cauchy a.s. if and only if

$$\lim_{n \to \infty} \mathbb{P}\left\{ \bigcup_{m=1}^{\infty} (|X_{m+n} - X_n| \geq \varepsilon) \right\} = 0 \quad \text{for every } \varepsilon > 0.$$

Corollary 0.13 *The sequence of random variables $\{X_n\}$ converges a.s. to the random variable X if*

$$\sum_{n=1}^{\infty} \mathbb{P}\{|X_n - X| \geq \varepsilon\} < \infty \quad for\ all\ \varepsilon > 0.$$

Definition 0.13 *(cf. Hsu and Robbins [49]) A sequence of random variables $\{X_n\}$ is said to converge completely to a random variable X if*

$$\sum_{n=1}^{\infty} \mathbb{P}\{|X_n - X| \geq \varepsilon\} < \infty \quad for\ every\ \varepsilon > 0.$$

According to Corollary 0.13, completely convergence implies a.s. convergence.

Corollary 0.14 *Let $\{X_n\}$ be a sequence of random variables, and let X be a random variable with $\mathbb{E}(X_n - X)^2 < \infty$, $n \geq 1$, such that $\sum_{n=1}^{\infty} \mathbb{E}(X_n - X)^2 < \infty$. Then $X_n \overset{a.s.}{\to} X$.*

A somewhat weaker concept of convergence is that of convergence in probability, which is identical with the concept of convergence in measure.

Definition 0.14 *Let $\{X_n, n \geq 1\}$ be a sequence of random variables defined on a probability space $(\Omega, \mathcal{F}, \mathbb{P})$. We say that $\{X_n\}$ converges in probability to the random variable X if for every $\varepsilon > 0$,*

$$\mathbb{P}\{\omega \in \Omega : |X_n(\omega) - X(\omega)| \geq \varepsilon\} \to 0$$

as $n \to \infty$. In this case we write $X_n \overset{\mathbb{P}}{\to} X$.

Definition 0.15 *We say that $\{X_n\}$ is Cauchy in probability (or fundamental in probability) if for every $\varepsilon > 0$,*

$$\mathbb{P}\{|X_n - X_m| \geq \varepsilon\} \to 0$$

as $m, n \to \infty$.

Remark 0.15 *If $X_n \overset{\mathbb{P}}{\to} X$, then X is unique a.s. in the sense that if $X_n \overset{\mathbb{P}}{\to} X$ and $X_n \overset{\mathbb{P}}{\to} Y$, the $X = Y$ a.s.*

Remark 0.16 *If $X_n \overset{\mathbb{P}}{\to} X$, then $\{X_n\}$ is Cauchy in probability.*

Remark 0.17 *If $X_n \overset{\mathbb{P}}{\to} X$, and let $\{X_{n_k}\}$ be any infinite subsequence of $\{X_n\}$. Then $X_{n_k} \overset{\mathbb{P}}{\to} X$ as $k \to \infty$.*

Proposition 0.24 *We have*

$$X_n \xrightarrow{a.s.} X \Rightarrow X_n \xrightarrow{\mathbb{P}} X.$$

Remark 0.18 *That the converse statement of Proposition 0.24 is not true in general. In some special case, however, convergence in probability implies a.s. convergence.*

Proposition 0.25 *Let $\{X_n\}$ be a sequence of random variables which converges in probability to the random variable X. Then there exists a subsequence $\{X_{n_k}\} \subset \{X_n\}$ which converges a.s. to X.*

Proposition 0.26 *The sequence $\{X_n\}$ converges in probability to a random variable X if and only if it is Cauchy in probability.*

Definition 0.16 *Let $\{X_n\}$ be a sequence of random variables such that $X_n \in \mathbb{L}^1$, $n \geq 1$. We say that $\{X_n\}$ converges in mean to a random variable $X \in \mathbb{L}^1$ if $\mathbb{E}|X_n - X| \to 0$ as $n \to \infty$. In this case we write $X_n \xrightarrow{\mathbb{L}^1} X$ and note that the limiting random variable X, if it exists, is unique a.s.*

Definition 0.17 *The sequence $\{X_n\}$ is said to be Cauchy in mean if*

$$\mathbb{E}|X_n - X_m| \to 0$$

as $m, n \to \infty$.

First we recall from measure theory the following three important results on convergence in mean.

Theorem 0.19 (*Lebesgue Dominated Convergence Theorem*) *Let $\{X_n\}$ be a sequence of random variables such that $\{X_n\}$ converges in probability (or a.s.) to a random variable X. Suppose there exists a random variable $Y \in \mathbb{L}^1$ such that $|X_n| \leq |Y|$ a.s. Then $X_n, X \in \mathbb{L}^1$, and, moreover, $X_n \xrightarrow{\mathbb{L}^1} X$. In particular, $\mathbb{E}X_n \to \mathbb{E}X$ as $n \to \infty$.*

Theorem 0.20 (*Monotone Convergence Theorem*) *Let $\{X_n\}$ be a nondecreasing (a.s.) sequence of nonnegative (a.s.) random variables which converges a.s. to a random variable X.*
(1) Suppose that $X_n \in \mathbb{L}^1$ for all $n \geq 1$ and $\lim_{n \to \infty} \mathbb{E}X_n < \infty$. Then $X \in \mathbb{L}^1$ and $\lim_{n \to \infty} \mathbb{E}X_n = \mathbb{E}X$.
(2) Conversely, if $X \in \mathbb{L}^1$, then each $X_n \in \mathbb{L}^1$ and $\lim_{n \to \infty} \mathbb{E}X_n = \mathbb{E}X$.

Theorem 0.21 (*Fatou's Lemma*) *Let $X_n \in \mathbb{L}^1$, $n \geq 1$, be a sequence of (a.s.) nonnegative random variables such that $\liminf_{n \to \infty} \mathbb{E}X_n < \infty$. Then*

$$X_* = \liminf_{n \to \infty} X_n \in \mathbb{L}^1,$$

and the inequality

$$\mathbb{E}X_* \leq \liminf_{n\to\infty} \mathbb{E}X_n$$

holds.

We next study some properties of convergence in mean.

Proposition 0.27 *We have the following:*
(a) $X_n \xrightarrow{\mathbb{L}^1} X \Rightarrow X_n \xrightarrow{\mathbb{P}} X$.
(b) $\{X_n\}$ *is Cauchy in mean* $\Rightarrow$ $\{X_n\}$ *is Cauchy in probability.*

Proposition 0.28 *The sequence* $\{X_n\}$, $X_n \in \mathbb{L}^1$, *is Cauchy in mean if and only if it converges in mean to some random variable* $X \in \mathbb{L}^1$.

Theorem 0.22 *In the set* $\mathbb{L}^1$, *define* $X = 0$ *if and only if* $X = 0$ *a.s. Then* $\mathbb{L}^1$ *is a (real) Banach space with respect to the norm* $\|X\|_1 = \mathbb{E}|X|$ *for* $X \in \mathbb{L}^1$.

Remark 0.23 *Let* X *be a random variable. Then* $\mathbb{E}|X| < \infty$ *if and only if there exists a sequence* $\{X_n\}$ *of random variables in* $\mathbb{L}^1$ *which is Cauchy in mean and which converges to* X *in probability. Moreover, in that case* $X_n \xrightarrow{\mathbb{L}^1} X$.

Remark 0.24 *Let* $X \in \mathbb{L}^1$. *Then for every* $\varepsilon > 0$ *there exists a simple random variables (simple function)* $Y \in \mathbb{L}^1$ *such that* $\mathbb{E}|X - Y| < \varepsilon$.

Finally we give the following necessary and sufficient condition for convergence in mean.

Proposition 0.29 *Let* $X_n \in \mathbb{L}^1$, $n = 1, 2, ...$ *Then the sequence* $\{X_n\}$ *converges in mean to some random variable* $X \in \mathbb{L}^1$ *if and only if the following two conditions are satisfied:*
(a) $X_n \xrightarrow{\mathbb{P}} X$.
(b) For every $\varepsilon > 0$, *there exists a* $\delta = \delta(\varepsilon) > 0$ *(independent of n) such that the inequality*

$$\int_E |X_n|d\mathbb{P} < \varepsilon$$

holds for every event E *for which* $\mathbb{P}(E) < \delta$ *and for all* $n \geq 1$.

Let $(\Omega, \mathcal{F}, \mathbb{P})$ be a probability space, and let $1 \leq p < \infty$. Denote by $\mathbb{L}^p = \mathbb{L}^p(\Omega, \mathcal{F}, \mathbb{P})$ the set of all random variables X on Ω such that $\mathbb{E}|X|^p < \infty$.

Definition 0.18 *Let* $X_n \in \mathbb{L}^p$, $n = 1, 2, ...$ *We say that the sequence* $\{X_n\}$ *converges in the pth mean to a random variable* $X \in \mathbb{L}^p$ *if*

$$\mathbb{E}|X_n - X|^p \to 0$$

as $n \to \infty$. *In this case write* $X_n \xrightarrow{\mathbb{L}^p} X$, *and note that the limiting random variable* X, *if it exists, is unique a.s.*

Definition 0.19 *The sequence $\{X_n\}$ is said to be Cauchy in the pth mean if*

$$\mathbb{E}|X_n - X_m|^p \to 0$$

as $m, n \to \infty$.

Clearly $X_n \overset{\mathrm{L}^p}{\to} X \Rightarrow X_n \overset{\mathbb{P}}{\to} X$ and $\{X_n\}$ Cauchy in the pth mean $\Rightarrow \{X_n\}$ Cauchy in probability.

Proposition 0.30 *The sequence $\{X_n\}$, $X_n \in \mathrm{L}^p$, is Cauchy in the pth mean if and only if it converges in the pth mean to some random variable $X \in \mathrm{L}^p$.*

Theorem 0.25 *Let $1 \le p < \infty$. In the set L^p, define the null vector $X = 0$ if and only if $X = 0$ a.s. Then L^p is a (real) Banach space with respect to the norm $\|X\|_p = [\mathbb{E}|X|^p]^{1/p}$ for $X \in \mathrm{L}^p$.*

Remark 0.26 *For the case $p = 1$, Proposition 0.30 provides an alternative proof of Proposition 0.28.*

Remark 0.27 *For the case $p = 2$, L^2 is a real Hilbert space with respect to the inner product $\langle X, Y \rangle = \mathbb{E}(XY)$, $X, Y \in \mathrm{L}^2$.*

Remark 0.28 *Let $1 \le p < \infty$, and let $X \in \mathrm{L}^p$. Let $\varepsilon > 0$. Then there exists a simple random variable $Y \in \mathrm{L}^p$ such that $\mathbb{E}|X - Y|^p < \varepsilon$.*

Let $\{F_n\}$ be a sequence of distribution functions. We recall that $\{F_n\}$ is a non-decreasing, right-continuous function which satisfies $F_n(-\infty) = 0$ and $F_n(+\infty) = 1$. To investigate in detail the convergence properties of the sequence $\{F_n\}$, it is necessary to introduce the concept of weak convergenvce in the general framework of bounded, nondecreasing, right-continuous functions. We note that such functions can have at most a countable number of discontinuity points (Proposition 0.2).

Definition 0.20 *Let $\{F_n\}$ be a sequence of uniformly bounded, nondecreasing, right-continuous functions defined on $\mathbb{R}$. We say that $\{F_n\}$ converges weakly to a bounded, nondecreasing, right-continuous functions F on $\mathbb{R}$ if*

$$F_n(x) \to F(x) \quad as \ n \to \infty$$

at all continuity points x of F. In this case we write $F_n \overset{w}{\to} F$.

Remark 0.29 *We note that the weak limit of the sequence $\{F_n\}$, if it exists, is unique. In fact, if $F_n \overset{w}{\to} F$ and $F_n \overset{w}{\to} F'$, then $F = F'$ on the set $C_F \cap C_{F'}$, where C_F and $C_{F'}$ are, respectively, the sets of continuity points of F and F'. From Proposition 0.3 it follows that $F = F'$.*

Definition 0.21 *Let* $\{F_n\}$ *be as in Definition 0.20. Then* $\{F_n\}$ *is said to converge completely to* F *if* $F_n \overset{w}{\to} F$ *and* $F_n(\mp\infty) \to F(\mp\infty)$ *as* $n \to \infty$. *In this case we write* $F_n \overset{c}{\to} F$.

Remark 0.30 *Let* $\{F_n\}$ *and* $\{F\}$ *be as in Definition 0.20, and suppose that* $F_n \overset{w}{\to}$ F. *Then* $F_n \overset{c}{\to} F$ *if and only if*

$$\sup_{n \geq 1}\{F_n(+\infty) - F_n(-\infty) - [F_n(a) - F_n(-a)]\} \to 0 \ \ as \ a \to \infty$$

or

$$F_n(+\infty) - F_n(-\infty) \to F(+\infty) - F(-\infty) \ \ as \ a \to \infty.$$

In cases where $\{F_n\}$ and F are distribution functions such that $F_n \overset{c}{\to} F$, let X_n and X be random variables corresponding to F_n and F, respectively. Then we say that X_n converges in law to X, and write $X_n \overset{L}{\to} X$.

Theorem 0.31 *(Kolmogorov) Let* $\{X_n\}$ *be a sequence of random variables such that* $X_n \overset{\mathbb{P}}{\to} X$. *Then* $X_n \overset{L}{\to} X$.

Corollary 0.15 *Let* $\{X_n\}$ *be a sequence of random variables, and* c *be a constant. Then*

$$X_n \overset{L}{\to} c \Leftrightarrow X_n \overset{\mathbb{P}}{\to} c.$$

Next we give the weak compactness criterion due to Helly.

Theorem 0.32 *(Helly's Theorem) Let* $\{F_n\}$ *be a sequence of uniformly bounded, nondecreasing, right-continuous functions. Then* $\{F_n\}$ *contains a subsequence* $\{F_{n_k}\}$ *which converges weakly to a bounded, nondecreasing, right-continuous functions.*

0.4   Martingale

We now consider a special case of dependence which has its origin in gambling.

Let $(\Omega, \mathcal{F}, \mathbb{P})$ be a probability space, and let $(T, \prec)$ be a partially ordered set. Let $\{\mathcal{D}_t, \ t \in T\}$ be a collection of sub-σ-fields of $\mathcal{F}$ such that $\mathcal{D}_s \subset \mathcal{D}_t$ for $s \prec t$, $s, \ t \in T$. Let $\mathcal{M} = \{X_t, \ t \in T\}$ be a collection of random variables defined on Ω, each having a finite expectation.

Definition 0.22 *The class* $\mathcal{M}$ *is said to constitute a martingale with respect to* $\{\mathcal{D}_t, \ t \in T\}$ *if the following conditions hold:*
(a) For every $t \in T$, X_t *is* $\mathcal{D}_t$-*measurable.*
(b) For $s, \ t \in T$, $s \prec t$, *the relation*

$$\mathbb{E}[X_t|\mathcal{D}_s] = X_s \quad a.s. \tag{0.28}$$

holds.

The class $\mathcal{M}$ is said to be a submartingale (respectively, supermartingale) if in (0.28) we replace the $=$ sign by the $\geq$ (respectively, $\leq$) sign.

Clearly, changing X_t into $-X_t$ interchanges "submartingale" and "supermartingale".

In the following, if $\{X_t\}$ is a martingale with respect to $\mathcal{D}_t$ (in the sense of Definition 0.22), we will say that $\{(X_t, \mathcal{D}_t)\}_{t \in T}$ or, simply, that $\{(X_t, \mathcal{D}_t)\}$ is a martingale. We will restrict attention mainly to the discrete parameter case where $T = \{1, 2, \ldots\}$ is the set of positive integers. In particular, if $\mathcal{D}_n = \sigma(X_1, X_2, \ldots, X_n)$ for $n \geq 1$, we will simply say that $\{X_n\}$ is a martingale.

Proposition 0.31 $\{(X_n, \mathcal{D}_n)\}_{n \geq 1}$ *is a martingale (submartingale or supermartingale) if and only if for every $n > 1$,*

$$\mathbb{E}[X_n | \mathcal{D}_{n-1}] = (\geq \text{ or } \leq) X_{n-1} \quad a.s. \tag{0.29}$$

Example 0.14 *Consider a sequence of independent, identically distributed (i.i.d.) random variables $\{X_n\}$ with common distribution given by $\mathbb{P}\{X_i = \pm 1\} = \frac{1}{2}$. We can interpret $X_i = 1$ as the event that a gambler who is playing a sequence of games in each of which he (or she) has probability $\frac{1}{2}$ of winning \$1 and probability $\frac{1}{2}$ of losing \$1 wins the ith game. Similarly, $X_i = -1$ is the event that he loses the ith game. If he wagers an amount $\$b_{i-1}$ on the ith game, his expected winnings on the ith game are*

$$b_{i-1}\mathbb{P}\{X_i = 1\} - b_{i-1}\mathbb{P}\{X_i = -1\} = 0.$$

Suppose that the gambler's strategy is to bet $b_0 (> 0)$ on the first (trial) game and $b_n = b_n(X_1, \ldots, X_n)$ on the $(n+1)$st game, $n \geq 1$, where b_n is $\mathcal{D}_n$-measurable. Let $S_0 (> b_0)$ be his initial fortune, and let

$$S_n = S_n(b_0, b_1, \ldots, b_{n-1}, X_1, \ldots, X_n)$$

be his fortune after n trials. Clearly

$$S_{n+1} = S_0 + \sum_{k=0}^{n} b_k X_{k+1} = S_n + X_{n+1} b_n.$$

Let $\mathcal{D}_n = \sigma(X_1, \ldots, X_n)$. Since $\mathbb{E}X_n = 0$ for $n \geq 1$, we have

$$\begin{aligned}
\mathbb{E}[S_{n+1} | \mathcal{D}_n] &= S_n + \mathbb{E}[X_{n+1} b_n(X_1, \ldots, X_n) | \mathcal{D}_n] \\
&= S_n + b_n \mathbb{E}X_{n+1} \\
&= S_n \quad a.s.,
\end{aligned}$$

so that $\{S_n, \mathcal{D}_n\}$ is a martingale. In particular, if $b_n(X_1, \ldots, X_n) \equiv 1$ for $n \geq 0$ and $S_0 = 0$, then $S_n = \sum_{k=1}^{n} X_k$ for $n \geq 1$.

Consider now the following strategy for the gambler. He doubles his bet until he wins a game and then quits. Clearly the probability that he will win at least one game equals $\sum_{k=1}^{\infty}(\frac{1}{2})^k = 1$, so that he is certain to win at least one game. Let $b_0 = b > 0$ be a constant. Then

$$b_{n-1} = \begin{cases} b2^{n-1} & \text{if } X_i = -1 \text{ for } i = 1, 2, \ldots, n-1, \\ 0 & \text{otherwise.} \end{cases}$$

If the gambler wins the first time on the $(n+1)$st game, he will have lost

$$\sum_{k=1}^{n} b2^{k-1} = b(2^n - 1)$$

on the first n games, and since he (bets and) wins $b2^n$ on the $(n+1)$st game, the probability is 1 that he will win b. The catch, of course, is that the gambler must have infinite initial fortune.

Finally, suppose that the gambler has the option of skipping individual games. Let $\delta_n = \delta_n(X_1, \ldots, X_{n-1})$ be a $\mathcal{D}_{n-1}$-measurable function taking the values

$$\delta_n = \begin{cases} 0 & \text{if he skips the nth game,} \\ 1 & \text{if he bets on the nth game.} \end{cases}$$

If S_n^* is his fortune after n trial, then

$$S_{n+1}^* = S_n^* + \delta_{n+1}(X_1, \ldots, X_n)b_n(X_1, \ldots, X_n)X_{n+1}.$$

It is easily seen that $\mathbb{E}|S_n^*| < \infty$ for all n, so that

$$\mathbb{E}[S_{n+1}^*|\mathcal{D}_n] = S_n^* + \delta_{n+1}b_n\mathbb{E}X_{n+1} = S_n^* \quad a.s.,$$

and $\{(S_n^*, \mathcal{D}_n)\}$ is a martingale.

Example 0.15 Let $\{X_n\}$ be a sequence of independent random variables such that $\mathbb{E}X_n = 0$ for all $n \geq 1$. Let $S_n = \sum_{k=1}^{n} X_k$, $n \geq 1$. Clearly

$$\mathbb{E}|S_n| \leq \sum_{k=1}^{n}\mathbb{E}|X_k| < \infty$$

for all n. Then $\{S_n\}$ is a martingale. In fact, for every $n \geq 1$ we have

$$\begin{aligned} \mathbb{E}[S_{n+1}|S_1, \ldots, S_n] &= \mathbb{E}[(S_n + X_{n+1})|X_1, \ldots, X_n] \\ &= S_n + \mathbb{E}X_{n+1} \\ &= S_n \quad a.s. \end{aligned}$$

Clearly it is sufficient to assume that $\mathbb{E}[X_{n+1}|X_1, \ldots, X_n] = 0$ a.s. for all $n \geq 1$, and independence of the X_n is not required. Also, if $\{S_n\}$ is a martingale sequence, then, by setting $X_n = S_n - S_{n-1}$ for $n \geq 1$, $S_0 = 0$, we have

$$\begin{aligned} \mathbb{E}[X_{n+1}|X_1, \ldots, X_n] &= \mathbb{E}[(S_n - S_{n-1})|X_1, \ldots, X_n] \\ &= S_n - S_n \\ &= 0 \quad a.s. \end{aligned}$$

It follows that the martingale property characterizes sums of random variables centered at conditional expectations, given the predecessors.

Example 0.16 Let X_1, X_2, $\ldots$ be independent random variables with $\mathbb{E}X_n = 1$ for all n. Let $Z_n = \prod_{i=1}^n X_i$, $n \geq 1$. Then $\mathbb{E}|Z_n| = \prod_{i=1}^n \mathbb{E}|X_i| < \infty$ for every n, and

$$\begin{aligned} \mathbb{E}[Z_{n+1}|Z_1, \ldots, Z_n] &= \mathbb{E}[(X_{n+1}Z_n)|Z_1, \ldots, Z_n] \\ &= Z_n \mathbb{E}[X_{n+1}|Z_1, \ldots, Z_n] \\ &= Z_n \quad a.s., \end{aligned}$$

so that Z_n is a martingale.

Example 0.17 *Consider an urn which contains $b \geq 1$ black and $w \geq 1$ white balls which are well mixed. Repeated drawings are made from the urn, and after each drawing the ball drawn is replaced, along with c balls of the same color. Here $c \geq 1$ is an integer. Let $X_0 = b/(b+w)$, and X_n be the proportion of black balls in the urn after the nth draw. We show that $\{X_n, n \geq 0\}$ is a martingale.*

Let $Y_0 = 1$, and for $n \geq 1$ let $Y_n = 1$ if the nth ball drawn is black, and $Y_n = 0$ if the nth ball drawn is white. Let b_n and w_n denote the number of black and white balls, respectively, in the urn after the nth draw. Write $b_0 = b$ and $w_0 = w$. Then $X_n = b_n/(b_n + w_n)$, $n \geq 0$. Clearly, for $n \geq 0$

$$b_{n+1} = b_n + cY_{n+1}, \quad w_{n+1} = w_n + c(1 - Y_{n+1}).$$

Now $\mathbb{P}\{Y_{n+1} = 1|Y_0, \ldots, Y_n\} = X_n$ for $n \geq 0$, so that

$$\begin{aligned} \mathbb{E}[X_{n+1}|Y_0, \ldots, Y_n] &= \mathbb{E}\left[\frac{b_{n+1}}{b_{n+1} + w_{n+1}}\bigg| Y_0, \ldots, Y_n\right] \\ &= \frac{b_{n+1}}{b_{n+1} + w_{n+1} + c} + \frac{b_{n+1}}{b_{n+1} + w_{n+1} + c}\mathbb{E}[Y_{n+1}|Y_0, \ldots, Y_n] \\ &= \frac{b_{n+1}}{b_{n+1} + w_{n+1} + c} + \frac{b_{n+1}}{b_{n+1} + w_{n+1} + c}X_n \\ &= \frac{b_n}{b_n + w_n} \\ &= X_n \quad a.s. \end{aligned}$$

Since $\sigma(X_0, \ldots, X_n) \subset \sigma(Y_0, \ldots, Y_n)$, *we see that*

$$E[X_{n+1}|X_0, \ldots, X_n] = E\big[E[X_{n+1}|Y_0, \ldots, Y_n]\big|X_0, \ldots, X_n\big]$$
$$= X_n \quad a.s.$$

It follows that $\{X_n\}$ *is a martingale. It also follows that*

$$\mathbb{P}\{Y_{n+1} = 1\} = E[\mathbb{P}\{Y_{n+1} = 1\}|Y_0, \ldots, Y_n]$$
$$= EX_n$$
$$= EX_0$$
$$= \frac{b}{b+w}.$$

Example 0.18 *Let* $\{X_n\}$ *be a sequence of random variables, and suppose that the joint probability density of* $(X_1, X_2, \ldots, X_n)$ *is either* p_n *or* q_n. *In statistics the ratio* $\lambda_n = \lambda_n(X_1, X_2, \ldots, X_n) = \frac{q_n(X_1, X_2, \ldots, X_n)}{p_n(X_1, X_2, \ldots, X_n)}$ *is known as a likelihood ratio. The ratio* λ_n *is likely to be small or large, according as the true probability density function is* p_n *or* q_n, *so that* λ_n *may be used to reach a decision.*

For convenience let us assume that $p_n > 0$ *and is continuous for all* n. *(It is sufficient to assume that* $q_n = 0$ *whenever* $p_n = 0$.) *If* p_n *is the true density, the conditional probablity density function of* X_{n+1}, *given* $X_1, X_2, \ldots, X_n$ *is* p_{n+1}/p_n, *so that*

$$E[\lambda_{n+1}|X_1 = x_1, \ldots, X_n = x_n]$$
$$= \int_{-\infty}^{\infty} \lambda_{n+1}(x_1, \ldots, x_n, y)\frac{p_{n+1}(x_1, \ldots, x_n, y)}{p_n(x_1, \ldots, x_n)}dy$$
$$= \int_{-\infty}^{\infty} \frac{q_{n+1}(x_1, \ldots, x_n, y)}{p_n(x_1, \ldots, x_n)}dy$$
$$= \frac{q_n(x_1, \ldots, x_n)}{p_n(x_1, \ldots, x_n)} \quad a.s.$$

It follows that $E[\lambda_{n+1}|X_1, \ldots, X_n] = \lambda_n$ *a.s., and hence*

$$E\big[E[\lambda_{n+1}|X_1, \ldots, X_n]\big|\lambda_1, \ldots, \lambda_n\big] = E[\lambda_{n+1}|\lambda_1, \ldots, \lambda_n] = \lambda_n \quad a.s.$$

in view of the fact that $\sigma(X_1, \ldots, X_n) \subset \sigma(\lambda_1, \ldots, \lambda_n)$. *Thus* $\{\lambda_n\}$ *is a martingale.*

The following proposition gives a procedure by which a martingale may be constructed.

Proposition 0.32 *Let* $\{\mathcal{D}_n\}$ *be a nondecreasing sequence of sub-σ-fields of* $\mathcal{F}$, *and let* X *be a random variable such that* $E|X| < \infty$. *Then the sequence* $X_n = E[X|\mathcal{D}_n]$ *a.s.,* $n \geq 1$, *is a martingale.*

Remark 0.33 *Given a martingale $\{X_n\}$, there does not necessarily exist a random variable X with $\mathbb{E}|X| < \infty$ and a sequence $\{\mathcal{D}_n\}$ of nondecreasing sub-σ-fields of $\mathcal{F}$ such that $X_n = \mathbb{E}[X|\mathcal{D}_n]$ a.s. Take $Z_1, Z_2, \ldots$ to be i.i.d. with distribution $\mathbb{P}\{Z_i = 0\} = \frac{1}{2} = \mathbb{P}\{Z_i = 2\}$. In Example 0.16 we showed that $X_n = \prod_{i=1}^{n} Z_i$ is a martingale sequence. Suppose there exists some random variable X with $\mathbb{E}|X| < \infty$ and sub-σ-fields $\mathcal{D}_n \subset \mathcal{D}_{n+1}$, $n \geq 1$, such that $X_n = \mathbb{E}[X|\mathcal{D}_n]$ a.s. Clearly X_n is $\mathcal{D}_n$-measurable, so that $A_n = \{X_n = 0\} \in \mathcal{D}_n$. By the definition of conditional probability*

$$0 = \int_{A_n} X_n d\mathbb{P}_{\mathcal{D}_n} = \int_{A_n} X d\mathbb{P} = \mathbb{E}(X \mathbf{1}_{A_n}).$$

Since $\lim_{n \to \infty} \mathbf{1}_{A_n} = 1$ a.s., it follows from the dominated convergence theorem that as $n \to \infty$,

$$0 = \int_{A_n} X_n d\mathbb{P}_{\mathcal{D}_n} = \mathbb{E}(X \mathbf{1}_{A_n}) \to \mathbb{E}X = \mathbb{E}X_n = 1,$$

which is a contradiction.

Proposition 0.33 *Let $\{X_n\}$ be a sequence of random variables with $\mathbb{E}|X_n| < \infty$ for all n. Then $\{X_n\}$ is a martingale (submartingale, supermartingale) if and only if, for every $m \geq n$ and $A \in \sigma(X_1, X_2, \ldots, X_n)$,*

$$\int_A X_m d\mathbb{P} = (\geq, \leq) \int_A X_n d\mathbb{P} \tag{0.30}$$

Proposition 0.34 *(Decomposition of Submartingales) Let $\{(X_n, \mathcal{D}_n)\}_{n \geq 1}$ be a submartingale. Then X_n has a decomposition*

$$X_n = X'_n + X''_n \quad a.s.,$$

where $\{(X'_n, \mathcal{D}_n)\}_{n \geq 1}$ is a martingale, and $\{X''_n\}$ is a nondecreasing sequence of a.s. nonnegative random variables such that X''_n is $\mathcal{D}_{n-1}$-measurable, $n \geq 2$.

Remark 0.34 *Let $X_n = X'_n + X''_n$, $n \geq 1$, where X'_n and X''_n are as in Proposition 0.34. Then for $n \geq 2$,*

$$\mathbb{E}[X_n|\mathcal{D}_{n-1}] = \mathbb{E}[X'_n|\mathcal{D}_{n-1}] + \mathbb{E}[X''_n|\mathcal{D}_{n-1}]$$
$$= X'_{n-1} + X''_n$$
$$\geq X_{n-1} \quad a.s.,$$

so that $\{X_n\}$ is a submartingale and the converse of Proposition 0.34 also holds.

Remark 0.35 *Let $\{X_n\}$ be a submartingale. In view of Proposition 0.34 we have*

$$\mathbb{E}X_n = \mathbb{E}X'_n + \mathbb{E}X''_n,$$

$$\mathbb{E}|X_n'| \leq \mathbb{E}|X_n| + \mathbb{E}|X_n''|,$$

and

$$0 \leq \mathbb{E}X_n'' \uparrow.$$

If $\sup_{n \geq 1} \mathbb{E}|X_n| < \infty$, *then* $\mathbb{E}X_n'$ *and* $\mathbb{E}X_n''$ *exist and*

$$\sup_{n \geq 1} \mathbb{E}|X_n'| < \infty, \quad \sup_{n \geq 1} \mathbb{E}X_n'' < \infty.$$

It follows that $0 \leq X_n'' \uparrow X''$ *a.s., and study of the convergence of the sequence* $\{X_n\}$ *reduces to that of the sequence* $\{X_n'\}$, *which is a martingale sequence with* $\sup_{n \geq 1} \mathbb{E}|X_n'| < \infty$.

Proposition 0.35 *Let* $\{X_n\}$ *be a martingale, and let* g *be a convex function on* $\mathbb{R}$. *Then* $\{g(X_n)\}$ *be a martingale, provided that* $\mathbb{E}|g(X_n)| < \infty$, *for* $n \geq 1$.

If $\{X_n\}$ is a submartingale and g is a convex, nondecreasing function such that $\mathbb{E}|g(X_n)| < \infty$, then $\{g(X_n)\}$ is a submartingale.

Corollary 0.16 *If* $\{X_n\}$ *is a submartingale, then so also is the sequence*

$$\{ \max\{X_n, a\} : \ n \geq 1\}$$

for any $a \in \mathbb{R}$.

We now study some limit properties of a martingale. In particular we will state the martingale convergence theorem, which has many applications. We start with the following weaker version of the martingale convergence theorem.

Theorem 0.36 *Let* $\{S_n\}$ *be a martingale with* $\mathbb{E}S_n^2 < c < \infty$ *for all* $n \geq 1$. *Then there exists a random variable* S *such that* S_n *converges a.s. and in mean square to* S. *Moreover,* $\mathbb{E}S_n = \mathbb{E}S$ *for all* n.

Corollary 0.17 *Let* $\{X_n\}$ *be a sequence of random variables such that*

$$\mathbb{E}[X_n | X_1, \ \ldots, \ X_{n-1}] = 0 \quad a.s.$$

for all n. *Then* $\sum_{k=1}^{\infty} \mathbb{E}X_k^2 / k^2 < \infty$ *implies* $n^{-1} S_n \to 0$ *a.s.*

Corollary 0.18 (*cf. Teicher* [132]) *Let* $\{X_n\}$ *be a sequence of independent random variables with* $\mathbb{E}X_n = 0$ *and* $\mathbb{E}X_n^2 = \sigma_n^2$, $n = 1, 2, \ldots$. *If the following conditions hold:*
(a) $\sum_{n=2}^{\infty} n^{-4} \sigma_n^2 \sum_{i=1}^{n-1} \sigma_i^2 < \infty$,
(b) $n^{-2} \sum_{j=1}^{n} \sigma_j^2 \to 0$,
(c) $\sum_{n=1}^{\infty} \mathbb{P}\{|X_n| \geq a_n\} < \infty$ *for some sequence of constants* $a_n > 0$ *with* $\sum_{n=1}^{\infty} n^{-4} a_n^2 \sigma_n^2 < \infty$,
then $n^{-1} S_n \to 0$ *a.s.*

We next give an important inequality due to Doob which is basic to the proof of our main result. Let $\{X_n\}$ be a sequence of random variables. The sequence $\{X_n\}$ has a limit, finite or infinite, if and only if the number of its oscillations between any two (rational) numbers a, b, $a < b$, is finite (depending on a, b, and $\omega \in \Omega$). We obtain an estimate of the expected number of such oscillations for a submartingale or supermartingale sequence $\{X_n\}$.

Let a, $b \in \mathbb{R}$ with $a < b$. Let $\{X_n\}$ be any sequence of random variables. For each $\omega \in \Omega$, set

$$T_1(\omega) = \min\{n : \ X_n(\omega) \le a\},$$
$$T_2(\omega) = \min\{n : \ n > T_1(\omega), \ X_n(\omega) \ge b\},$$

$$\vdots$$

$$T_{2k-1}(\omega) = \min\{n : \ n > T_{2k-2}(\omega), \ X_n(\omega) \le a\},$$
$$T_{2k}(\omega) = \min\{n : \ n > T_{2k-1}(\omega), \ X_n(\omega) \ge b\},$$

$$\vdots$$

with the convention that if any set on the right-hand side is empty the corresponding $T_s(\omega) = +\infty$. Clearly $T_1 < T_2 < \cdots$ a.s., and T_s is extended (here T_s may take the value $+\infty$ with a positive probability) random variable for every s satisfying $T_s \ge s$ a.s.

For each $n \ge 1$ let us set $U_n(a,b)(\omega) = 0$ if $T_2(\omega) > n$, and $= \max\{s : T_{2s} \le n\}$ otherwise, and call $U_n(a,b)$ the number of upcrossings of the interval $[a, b]$ by $X_1, \ldots, X_n$. Clearly $U_n(a,b) \le [n/2]$, where $[x]$ is the largest integer $\le x$. Moreover

$$U_n(a,b) = \sum_{s=1}^{[n/2]} s 1_{\{T_{2s} \le n, \ T_{2s+2} > n\}},$$

so that $U_n(a,b)$ is a nonnegative, nondecreasing, and bounded random variable and hence is integrable.

Next, for $j \ge 2$, let

$$Y_j(\omega) = \begin{cases} 0 & \text{if for some } s, \ T_{2s+1}(\omega) < j \le T_{2s+2}(\omega), \\ 1 & \text{otherwise.} \end{cases}$$

Then

$$Y_2 = 1_{\{X_1 > a\}},$$

and for $j > 2$,

$$Y_j = 1_{\{Y_{j-1}=0, \ X_{j-1} \ge b\} \cup \{Y_{j-1}=1, \ X_{j-1} > a\}}.$$

It follows immediately that Y_j is a random variable and that Y_j is $\sigma(X_1, \ldots, X_{j-1})$-measurable.

Theorem 0.37 (*Doob's Upcrossing Inequality*) *Let* $\{X_1, X_2, \ldots, X_n\}$ *be a submartingale with respect to* $\mathcal{D}_1 \subset \mathcal{D}_2 \subset \cdots \subset \mathcal{D}_n \subset \mathcal{F}$. *Then*

$$EU_n(a, b) \leq \frac{\mathbb{E}(X_n - a)_+}{b - a}.$$

Theorem 0.38 (*Submartingale Convergence Theorem*) *Let* $\{(X_n, \mathcal{D}_n)\}_{n \geq 1}$ *be a submartingale. Suppose that* $\limsup_{n \to \infty} \mathbb{E}|X_n| < \infty$. *Then there exists a random variable* X *which is* $\sigma\left(\bigcup_{n=1}^{\infty} \mathcal{D}_n\right)$*-measurable such that* $X_n \to X$ *a.s. Moreover, the inequality*

$$\mathbb{E}|X| \leq \limsup_{n \to \infty} \mathbb{E}|X_n| < \infty$$

holds.

Remark 0.39 *Theorem 0.38 holds if the condition* $\limsup_{n \to \infty} \mathbb{E}|X_n| < \infty$ *is replaced by the condition* $\sup_{n \geq 1} \mathbb{E}|X_n| < \infty$. *In this case*

$$\mathbb{E}|X| \leq \sup_{n \geq 1} \mathbb{E}|X_n| < \infty.$$

Remark 0.40 *If* $\{X_n\}$ *is a supermartingale,* $\{-X_n\}$ *is a submartingale, so that 0.38 holds also when* $\{X_n\}$ *is a supermartingale. Since a martingale is also a submartingale, the result also holds for martingale.*

Remark 0.41 *Note that*

$$\limsup_{n \to \infty} \mathbb{E}|X_n| < \infty \Leftrightarrow \limsup_{n \to \infty} \mathbb{E}(X_n)_+ < \infty$$

and

$$\sup_{n \geq 1} \mathbb{E}|X_n| < \infty \Leftrightarrow \sup_{n \geq 1} \mathbb{E}(X_n)_+ < \infty.$$

There condition are satisfied, in particular, if $\{X_n\}$ *convergence in* $\mathbb{L}^1$. *In fact, since* $\{X_n\}$ *is a submartingale,*

$$\mathbb{E}|X_n| = 2\mathbb{E}(X_n)_+ - \mathbb{E}X_n \leq 2\mathbb{E}(X_n)_+ - \mathbb{E}X_1.$$

It follows that

$$\limsup_{n \to \infty} \mathbb{E}|X_n| \leq 2\limsup_{n \to \infty} \mathbb{E}(X_n)_+ - \mathbb{E}X_1$$

and

$$\sup_{n \geq 1} \mathbb{E}|X_n| \leq 2\sup_{n \geq 1} \mathbb{E}(X_n)_+ - \mathbb{E}X_1.$$

Conversely, $\mathbb{E}(X_n)_+ \leq \mathbb{E}|X_n|$, *so that*

$$\limsup_{n \to \infty} \mathbb{E}|X_n| < \infty \quad \left(\sup_{n \geq 1} \mathbb{E}|X_n| < \infty\right)$$

implies $\limsup_{n\to\infty} \mathbb{E}(X_n)_+ < \infty$ $(\sup_{n\geq 1} \mathbb{E}(X_n)_+ < \infty)$. *Note that, since* $\{X_n\}$
is a submartingale, so is $\{(X_n)_+\}$, *so that* $\{\mathbb{E}(X_n)_+\}$ *is a nondecreasing sequence*
of nonnegative real numbers and $\limsup_{n\to\infty} \mathbb{E}(X_n)_+$ $(\sup_{n\geq 1} \mathbb{E}(X_n)_+)$ *is either*
finite or $+\infty$.

Remark 0.42 *The condition* $\limsup_{n\to\infty} \mathbb{E}(X_n)_+ < \infty$ $(\sup_{n\geq 1} \mathbb{E}|X_n| < \infty)$ *of*
Theorem 0.38 is not sufficient to ensure that $X_n \to X$ *in* $\mathbb{L}^1$. *Let* Z_1, Z_2, $\ldots$
be i.i.d. random variables with $\mathbb{P}\{Z_n = 0\} = \mathbb{P}\{Z_n = 2\} = \frac{1}{2}$. *Let* $X_n =$
$\prod_{i=1}^{n} Z_i$, $n = 1$, 2, $\ldots$. *Then* $\{X_n\}$ *is a martingale such that* $\mathbb{E}X_n = 1$ *for*
all n. *It follows from Theorem 0.38 that* $X_n \to X$ *a.s., say. Clearly* $X = 0$ *a.s.*
However, $\mathbb{E}|X_n - X| = \mathbb{E}X_n = 1$ *for all* n, *so that* $\mathbb{E}|X_n - X| \not\to 0$.

Remark 0.43 *Every uniformly bounded submartingale (or supermartingale) con-*
verges a.s. Also every nonnegative martingale converges a.s., since

$$\limsup_{n\to\infty} \mathbb{E}|X_n| = \limsup_{n\to\infty} \mathbb{E}X_n = \mathbb{E}X_1 < \infty.$$

In fact, every positive supermartingale and every negative submartingale converge
a.s.

Let $(\mathbb{B}, \|\cdot\|)$ be a real separable Banach space. For any real number $p \geq 1$, de-
note by $\mathbb{L}^p_{\mathbb{B}}$ the space of $\mathbb{B}$-valued random elements such that $\|X\|_{\mathbb{L}^p_{\mathbb{B}}} = (\mathbb{E}\|X\|^p)^{1/p}$
is finite. Let $\mathcal{F}_0 = \{\emptyset, \Omega\} \subset \mathcal{F}_1 \subset \cdots$ be an increasing sequence of sub-σ-fields
of $\mathcal{F}$. Let $\{(X_j, \mathcal{F}_j)\}_{j=1}^{n}$ be an adapted sequence of $\mathbb{B}$-valued random elements
defined on $(\Omega, \mathcal{F}, \mathbb{P})$, that is, for every $j \geq 1$, X_j is $\mathcal{F}_j$ measurable. We call it a
sequence of $\mathbb{B}$-valued martingale differences if additionally $\mathbb{E}\left[X_j | \mathcal{F}_{j-1}\right] = 0$ a.s.
and X_j belongs to $\mathbb{L}^1_{\mathbb{B}}$ for any $j \geq 1$, and a sequence of $\mathbb{B}$-valued supermartingale
differences if additionally $\mathbb{E}\left[X_j | \mathcal{F}_{j-1}\right] \leq 0$ a.s. and X_j belongs to $\mathbb{L}^1_{\mathbb{B}}$ for any
$j \geq 1$.

0.5 Slowly varing function

Let ℓ be a positive measurable function, defined on some neighbourhood $[X, \infty)$
of infinity, and satisfying

$$\lim_{x\to\infty} \frac{\ell(\lambda x)}{\ell(x)} = 1 \quad \forall \lambda > 0, \tag{0.31}$$

then ℓ is said to be slowly varying (in Karamata's sense).

These function were introduced and studied by Karamata (1930, [65]) in a
pioneering paper, with continuity in place of measurability.

The neighbourhood $[X, \infty)$ is of little importance; we may (and often shall)
suppose ℓ defined on $(0, \infty)$–for instance, by taking $\ell(x) \equiv \ell(X)$ on $(0, X)$.

The basic Uniform Convergence Theorem—the most important theorem in the subject—was given by Karamata (1930, [65]) in the continuous case, Korevaar et al. (1949, [69]) in the measurable case.

Theorem 0.44 (*Uniform Convergence Theorem, or UCL*) *If ℓ is slowly varying, then*

$$\lim_{x \to \infty} \frac{\ell(\lambda x)}{\ell(x)} = 1, \quad \text{uniformly on each compact } \lambda\text{-set in } (0, \infty). \quad (0.32)$$

It is actually local uniformity that can be proved—uniformity in each compact subset of the set of λ.

We note at this point that measurability may be repalced by the Baire property.

Theorem 0.45 *If ℓ is Baire, then (0.31) implies (0.32).*

We turn now to the question of exactly which function ℓ can satisfy (0.31). The basic representation below is due to Karamata (1930, [65]) in the continuous case, Korevaar et al. (1949, [69]) in the measurable case. Following it, we give two further results, in which components of the representations used have many extra properties. We will be concerned with the measurable case except where indicated.

Theorem 0.46 (*Representation Theorem*) *The function ℓ is slowly varying if and only if it may be written in the form*

$$\ell(x) = c(x) \exp\left\{ \int_a^x \frac{\varepsilon(u)}{u} du \right\} \quad (x \ge a) \quad (0.33)$$

for some $a > 0$, where $c(\cdot)$ is measurable and $c(x) \to c \in (0, \infty)$, $\varepsilon(x) \to 0$ as $x \to \infty$.

Since ℓ, c, ε may be altered at will on finite intervals, the value of a is unimportant (one may choose to take $a = 1$, or $a = 0$ on taking $\varepsilon \equiv 0$ on a neighbourhood of 0 to avoid divergence of the integral at the origin), and one may also take c eventually bounded. We may re-write (0.33) as

$$\ell(x) = \exp\left\{ c_1(x) + \int_a^x \frac{\varepsilon(u)}{u} du \right\}, \quad (0.34)$$

where $c_1(x)$, $\varepsilon(x)$ are bounded and measurable, $c_1(x) \to d \in \mathbb{R}$, $\varepsilon(x) \to 0$ as $x \to \infty$.

We note an important property of functions satisfying (0.31).

Proposition 0.36 *If ℓ is positive, measurable, defined on some $[A, \infty)$, and*

$$\lim_{x \to \infty} \frac{\ell(\lambda x)}{\ell(x)} = 1 \quad \forall \lambda > 0,$$

then ℓ is bounded on all finite intervals far enough to the right. If $h(x) := \log \ell(e^x)$, h is also bounded on finite intervals far enough to the right.

We shall call a function locally bounded in some set A if it is bounded in each compact subset of A. Thus the above proposition concludes that there exists $X > 0$ such that ℓ is locally bounded in $[X, \infty)$. Since ℓ is measurable, it is also locally integrable in $[X, \infty)$, $\ell \in \mathbb{L}^1_{loc}[X, \infty)$.

Using the Representation Theorem, specific examples of slowly varying functions may be constructed at will. Trivially, (positive, measurable) functions with positive limits at infinity (in particular, positive constants) are slowly varying. Of course, the simplest non-trivial example is $\ell(x) = \log x$ $(c(x) \equiv 1,\ \varepsilon(x) \equiv \frac{1}{\log x})$. The iterates $\log \log x (= \log_2 x)$, $\log_k x (= \log \log_{k-1} x)$ are also slowly varying as are powers of $\log_k x$, rational functions with positive coefficients formed from the $\log_k x$, etc. Non-logarithmic examples are given by

$$\ell(x) = \exp\left\{(\log x)^{\alpha_1}(\log_2 x)^{\alpha_2}\ldots(\log_k x)^{\alpha_k}\right\} \quad (0 < \alpha_i < 1),$$

and

$$\ell(x) = \exp\left\{(\log x)/\log \log x\right\}.$$

Note that a slowly varying function ℓ may have

$$\liminf_{x \to \infty} \ell(x) = 0, \quad \limsup_{x \to \infty} \ell(x) = \infty$$

(that is, may exhibit "infinite oscillation"), an example being

$$\ell(x) = \exp\left\{(\log x)^{\frac{1}{3}} \cos(\log x)^{\frac{1}{3}}\right\}.$$

Finally, we note some elementary properties of slowly varying functions, whose proofs may be left to the reader.

Proposition 0.37
(a) If ℓ varies slowly, $\lim_{x \to \infty} \frac{\log \ell(x)}{\log x} = 0$.
(b) If ℓ varies slowly, so does $\left(\ell(x)\right)^\alpha$ for every $\alpha \in \mathbb{R}$.
(c) If ℓ_1, ℓ_2 vary slowly, so do $\ell_1(x)\ell_2(x)$, $\ell_1(x)+\ell_2(x)$, and $(if \lim_{x \to \infty} \ell_2(x) = \infty)$ $\ell_1(\ell_2(x))$.
(d) If $\ell_1, \ldots, \ell_k$ vary slowly and $r(x_1, \ldots, x_k)$ is a rational function with positive coefficients, $r(\ell_1(x), \ldots, \ell_k(x))$ varies slowly.
(e) If ℓ varies slowly and $\alpha > 0$,

$$\lim_{x \to \infty} x^\alpha \ell(x) = \infty, \quad \lim_{x \to \infty} x^{-\alpha}\ell(x) = 0.$$

A measurable [Baire] function $f > 0$ satisfying

$$\lim_{x \to \infty} \frac{f(\lambda x)}{f(x)} = \lambda^\rho \quad \forall \lambda > 0 \tag{0.35}$$

is called regularly varying [Baire regularly varying] of index ρ; we write $f \in R_\rho$ [$f \in BR_\rho$].

The definition and principal properties of regularly varying functions are due to Karamata (1930, [65]) in the case of continuous functions, Korevaar et al. (1949, [69]) for measurable functions; use of the Baire property in this context is due to Matuszewska (1965, [104]).

The next result concerns global bounds for $f(y)/f(x)$, extending Potter (1942, [111]):

Theorem 0.47 (*Potter's Theorem*)
(a) *If ℓ is slowly varying then for any chosen constants $A > 1$, $\delta > 0$ there exists $X = X(A, \delta)$ such that*

$$\ell(y)/\ell(x) \leq A \max\left\{(y/x)^\delta, \ (y/x)^{-\delta}\right\} \quad (x \geq X, \ y \geq X).$$

(b) *If, further, ℓ is bounded away from 0 and ∞ on every compact subset of $[0, \infty)$, then for every $\delta > 0$ there exists $A' = A'(\delta) > 1$ such that*

$$\ell(y)/\ell(x) \leq A' \max\left\{(y/x)^\delta, \ (y/x)^{-\delta}\right\} \quad (x > 0, \ y > 0).$$

(c) *If f is regularly varying of index ρ then for any chosen constants $A > 1$, $\delta > 0$ there exists $X = X(A, \delta)$ such that*

$$f(y)/f(x) \leq A \max\left\{(y/x)^{\rho+\delta}, \ (y/x)^{\rho-\delta}\right\} \quad (x \geq X, \ y \geq X).$$

The asymptotic behaviour of integrals of regularly varying functions will be of importance later.

Proposition 0.38 *If ℓ is slowly varying, X is so large that $\ell(x)$ is locally bounded in $[X, \infty)$, and $\alpha > -1$, then*

$$\int_X^x t^\alpha \ell(t)dt \sim x^{\alpha+1}\ell(x)/(\alpha + 1) \quad (x \to \infty).$$

The result remains true for $\alpha = 1$ in the sense that then

$$\frac{1}{\ell(x)} \int_X^x t^{-1}\ell(t)dt \to \infty, \tag{0.36}$$

but what is more important here is that the integral is now slowly, rather than regularly, varying.

Proposition 0.39 *Let ℓ be slowly varying and choose X so that $\ell \in \mathbb{L}^1_{loc}[X, \infty)$. Then $\int_X^x t^{-1}\ell(t)dt$ is slowly varying and (0.36) holds.*

With ℓ slowly varying, $\int^\infty t^{-1}\ell(t)dt$ may or may not converge. In the case of convergence, slow variation of $\int_X^x t^{-1}\ell(t)dt$ is trivial; a more interesting result is

Proposition 0.40 *If ℓ is slowly varying and*

$$\int^{\infty} t^{-1}\ell(t)dt < \infty,$$

then $\int_x^{\infty} t^{-1}\ell(t)dt$ is slowly varying, and

$$\frac{1}{\ell(x)} \int_x^{\infty} t^{-1}\ell(t)dt \to \infty \quad (x \to \infty). \tag{0.37}$$

There is also an analogue of Proposition 0.38 in which we integrate up to infinity:

Proposition 0.41 *If ℓ is slowly varying and $\alpha < -1$, then $\int^{\infty} t^{\alpha}\ell(t)dt$ converges and*

$$\frac{x^{\alpha+1}\ell(x)}{\int_x^{\infty} t^{\alpha}\ell(t)dt} \to -\alpha - 1 \quad (x \to \infty).$$

0.6   The laws of large numbers

In probability theory we study limit theorems which are of the following two types:
(a) Strong limit theorems. These deal with the a.s. convergence of a sequence of random variables.
(b) Weak limit theorems. These deal with the convergence in probability of a sequence of random variables, as well as convergence of a sequence of distribution functions.
In the section, for the most part, we shall give some important strong limit theorems of probability theory.
    Let $(\Omega, \mathcal{F}, \mathbb{P})$ be a probability space, and let $\{X_n\}$ be a sequence of random variables defined on it. The weak law of large numbers deals with conditions of convergence in probability of the sequence of partial sums $S_n = \sum_{k=1}^{n} X_k$, $n = 1, 2, \ldots$.

Theorem 0.48 (*Chebyshev's Weak Law of Large Number*) *Let $\{X_n\}$ be a sequence of independent random variables with $\mathbb{E}X_n^2 < \infty$, $n \geq 1$. Suppose there exists a $\gamma > 0$ such that $var(X_n) \leq \gamma$ for all $n \geq 1$. Set $S_n = \sum_{k=1}^{n} X_k$, $n \geq 1$. Then*

$$\frac{S_n - \mathbb{E}S_n}{n} \xrightarrow{\mathbb{P}} 0 \ \text{ as } n \to \infty.$$

Remark 0.49 *Let $\{X_n\}$ be a sequence of random variables, and $\{A_n\}$ be a sequence of constants. A statement of the type*

$$\frac{S_n - A_n}{n} \xrightarrow{\mathbb{P}} 0 \ \text{ as } n \to \infty$$

is called a weak law of large numbers. Sometimes one seeks sequences of constants $\{A_n\}$ and $\{B_n\}$, $B_n > 0$ and $B_n \to \infty$ as $n \to \infty$, such that

$$\frac{S_n - A_n}{B_n} \xrightarrow{\text{P}} 0 \quad as\ n \to \infty$$

Here we consider only the case $B_n = n$. In Theorem 0.51 we will show that, when the X_n are i.i.d. with common mean μ, the weak law of large numbers holds if we choose $A_n = n\mu$.

We next consider an important particular case of Theorem 0.48. For this purpose we need the following definition.

Definition 0.23 *Two random variables X and Y are said to be identically distributed if their probability distributions $\mathbb{P}_X$ and $\mathbb{P}_Y$ coincide on $\mathbb{B}$.*

Clearly X and Y are identically distributed if and only if their distribution functions coincide on $\mathbb{R}$. Moreover, in this case, for any real-valued Borel-measureable function g on $\mathbb{R}$ which is integrable with respect to $\mathbb{P}_X$ the relation

$$\mathbb{E}g(X) = \mathbb{E}g(Y)$$

holds.

Corollary 0.19 *Let $\{X_n\}$ be a sequence of i.i.d. random variables with common mean μ and variance $\sigma^2 < \infty$. Then*

$$\frac{S_n}{n} \xrightarrow{\text{P}} \mu \quad as\ n \to \infty.$$

We next give a necessary and sufficient condition for convergence in probability.

Theorem 0.50 *Let $\{Y_n\}$ be an arbitrary sequence of random variables. Then $Y_n \xrightarrow{\text{P}} 0$ if and only if*

$$\mathbb{E}\frac{|Y_n|^r}{1 + |Y_n|^r} \to 0 \quad for\ some\ r > 0.$$

Corollary 0.20 *Let $\{X_n\}$ be an arbitrary sequence of random variables. Then*

$$\frac{S_n - \mathbb{E}S_n}{n} \xrightarrow{\text{P}} 0 \quad as\ n \to \infty$$

if and only if

$$\mathbb{E}\frac{(S_n - \mathbb{E}S_n)^2}{n^2 + (S_n - \mathbb{E}S_n)^2} \to 0 \quad as\ n \to \infty.$$

We note that Theorem 0.48 and its corollary all follow from Theorem 0.50.

Finally we prove the following stronger result for the case of i.i.d. random variables, which is due to Khintchine.

Theorem 0.51 *Let $\{X_n\}$ be a sequence of i.i.d. random variables with common mean μ. Then*

$$\frac{S_n}{n} \xrightarrow{\mathbb{P}} \mu \ \ as \ n \to \infty.$$

Here we discuss some important tools of probability (limit) theory which we shall use in our subsequent investigation.

Proposition 0.42 *(Borel-Cantelli Lemma) Let $\{E_n\}$ be a sequence of events.*
(a) If $\Sigma_{n=1}^{\infty}\mathbb{P}(E_n) < \infty$, then $\mathbb{P}(\limsup_n E_n) = 0$.
(b) If, in addition, the E_n are independent, then $\mathbb{P}(\limsup_n E_n) = 0$ or $= 1$ according as the series $\Sigma_{n=1}^{\infty}\mathbb{P}(E_n)$ converges or diverges.

Corollary 0.21 *Let $\{E_n\}$ be an independent sequence of events. Then*

$$\mathbb{P}(\limsup_n E_n) = 0 \Leftrightarrow \Sigma_{n=1}^{\infty}\mathbb{P}(E_n) < \infty.$$

Corollary 0.22 *Let $\{X_n\}$ be a sequence of independent random variables. Then $X_n \xrightarrow{a.s.} 0$ if and only if $\Sigma_{n=1}^{\infty}\mathbb{P}\{|X_n| \geq \varepsilon\} < \infty$ for every $\varepsilon > 0$.*

Remark 0.52 *In view of the definition of the limit superior of a sequence of events, it is customary to write*

$$\limsup_n E_n = E_n \ \text{i.o.},$$

where i.o. *is an abbreviation for "infinitely often". Thus, if the E_n are independent events, $\mathbb{P}(E_n \ \text{i.o.})$ if and only if $\Sigma_{n=1}^{\infty}\mathbb{P}(E_n) < \infty$. Similarly, if $\{X_n\}$ is a sequence of independent random variables, $X_n \xrightarrow{a.s.} 0$ if and only if $\mathbb{P}\{|X_n| \geq \varepsilon \ \text{i.o.}\}$ for every $\varepsilon > 0$.*

Part (b) of Proposition 0.42 can be extended in several directions when the assumption of independence is dropped. Needless to say, some additional condition(s) will then be imposed. We consider one such extension in the following proposition.

Proposition 0.43 *Let $\{E_n\}$ be any sequence of events such that $\Sigma_{n=1}^{\infty}\mathbb{P}(E_n) = \infty$ and*

$$\liminf_{n\to\infty} \frac{\Sigma_{j=1}^{n}\Sigma_{k=1}^{n}\mathbb{P}(E_j \cap E_k)}{[\Sigma_{j=1}^{n}\mathbb{P}(E_j)]^2} = 1.$$

Then $\mathbb{P}(\limsup_n E_n) = 1$.

Corollary 0.23 *Let $\{E_n\}$ be pairwise independent, that is, suppose that $\mathbb{P}(E_j \cap E_k) = \mathbb{P}(E_j)\mathbb{P}(E_k)$ for all j, k, $j \neq k$. In addition, assume that $\sum_{n=1}^{\infty} \mathbb{P}(E_n) = \infty$. Then $\mathbb{P}(\limsup_n E_n) = 1$.*

We first give two elementary results on convergence of sequences of real numbers.

Lemma 0.1 (*Toeplitz Lemma*) *Let $\{a_n\}$ be a sequence of real numbers such that $a_n \to a$ as $n \to \infty$. Then $n^{-1}\sum_{k=1}^{n} a_k \to a$ as $n \to \infty$.*

Lemma 0.2 (*Kronecker Lemma*) *Let $\{a_n\}$ be a sequence of real numbers such that $\sum_{n=1}^{\infty} a_n$ converges. Then $n^{-1}\sum_{k=1}^{n} k a_k \to 0$ as $n \to \infty$.*

We next give the following result, due to Kolmogorov.

Proposition 0.44 *Let $\{X_n\}$ be a sequence of independent random variables with $var(X_n) = \sigma_n^2 < \infty$, $n = 1$, 2, ... Suppose that $\sum_{n=1}^{\infty}(\sigma_n^2/n^2) < \infty$. Let $S_n = \sum_{k=1}^{n} X_k$, $n = 1$, 2, ... Then the sequence $\{n^{-1}(S_n - \mathbb{E}S_n)\}$ converges a.s. to zero.*

We are now in a position to give the strong law of large numbers, for sequences of i.i.d. random variables.

Theorem 0.53 (*Kolmogorov*) *Let $\{X_n\}$ be a sequence of i.i.d. random variables. Let $S_n = \sum_{k=1}^{n} X_k$, $n = 1$, 2, ... Then the sequence $\{n^{-1}S_n\}$ converges a.s. to a finite limit α if and only if $\mathbb{E}|X_n| < \infty$. Moreover, in this case $\mathbb{E}X_n = \alpha$.*

0.7   The central limit theorems

We begin with the bounded variance case in the section. The case of most importance is that of the convergence in law of centered and normed partial sums to a normal random variable.

Let S_n be the number of successes in n independent trials with probability p of success in each trial, $0 < p < 1$. Then S_n is a binomial random variable with $\mathbb{E}S_n = np$ and $var(S_n) = np(1-p)$. DeMoivre and Laplace were the first to show that the sequence $\{(S_n - \mathbb{E}S_n)/\sqrt{var(S_n)}\}$ converges in law to the standard normal random variable as the number of trials n increases indefinitely. This result can clearly be reformulated as follows. Let $\{X_n\}$ be a sequence of i.i.d. random variables with $\mathbb{P}\{X_1 = 1\} = p$, $\mathbb{P}\{X_1 = 0\} = 1 - p$, $0 < p < 1$. Let $S_n = \sum_{k=1}^{n} X_k$, $n \geq 1$. Then

$$\frac{S_n - np}{\sqrt{np(1-p)}} \xrightarrow{L} X \quad \text{as } n \to \infty,$$

where X has the standard normal distribution. Motivated by this result, Lévy investigated the case of i.i.d. random variables with finite variance and obtained the most useful version of the celebrated central limit theorem. Later investigation centered around dropping the condition of identical distribution. In this direction Lindeberg obtained a set of sufficient conditions (which were later shown also to be necessary by Feller) for the convergence of suitably centered and normed S_n to the normal distribution.

We first give the classical result of DeMoivre and Laplace.

Proposition 0.45 *Let X_1, X_2, ... be a sequence of i.i.d. Bernoulli random variables with common distribution given by $\mathbb{P}\{X_j = 1\} = p$, $0 < p < 1$, and $\mathbb{P}\{X_j = 0\} = 1 - p = q$. Let $S_n = \Sigma_{j=1}^n X_j$, $n = 1$, 2, ... Then for every $x \in \mathbb{R}$,*

$$\lim_{n \to \infty} \mathbb{P}\left\{\frac{S_n - np}{\sqrt{npq}} \leq x\right\} = \frac{1}{\sqrt{2\pi}} \int_{-\infty}^{x} e^{-u^2/2} du.$$

We now consider the most general result in this direction for the case of independent but not necessarily identically distributed random variables.

Theorem 0.54 (*Lindeberg-Feller Central Limit Theorem*) *Let $\{X_n\}$ be a sequence of independent but not necessarily identically distributed random variables with $var(X_n) = \sigma_n^2 < \infty$, $n = 1$, 2, ... Let $\mathbb{E}X_n = \alpha_n$ and $S_n = \Sigma_{j=1}^n X_j$, $n = 1$, 2, ... Set $var(S_n) = B_n^2$. Let F_n be the distribution function of X_n. Then the following two conditions:*

(a) $\lim_{n \to \infty} \max_{1 \leq k \leq n}(\sigma_k^2/B_n^2) = 0$,

(b) $\lim_{n \to \infty} \mathbb{P}\left\{\frac{S_n - \mathbb{E}S_n}{B_n} \leq x\right\} = \frac{1}{\sqrt{2\pi}} \int_{-\infty}^{x} e^{-u^2/2} du$, for every $x \in \mathbb{R}$, hold if and only if for every $\varepsilon > 0$ the condition

$$\lim_{n \to \infty} \frac{1}{B_n^2} \sum_{k=1}^{n} \int_{|x - \alpha_k| \geq \varepsilon B_n} (x - \alpha_k)^2 dF_k(x) = 0 \tag{0.38}$$

is satisfied.

[Condition (0.38) is known as the Lindeberg condition. It should be noted that the convergence in (b) is uniform in x.]

Corollary 0.24 (*Lévy Central Limit Theorem*) *Let $\{X_n\}$ be a sequence of independent but not necessarily identically distributed random variables with $0 < var(X_n) = \sigma^2 < \infty$, $n = 1$, 2, ... Let $S_n = \Sigma_{k=1}^n X_k$, $n = 1$, 2, ... Then for every $x \in \mathbb{R}$,*

$$\lim_{n \to \infty} \mathbb{P}\left\{\frac{S_n - \mathbb{E}S_n}{\sigma\sqrt{n}} \leq x\right\} = \frac{1}{\sqrt{2\pi}} \int_{-\infty}^{x} e^{-u^2/2} du.$$

Corollary 0.25 (*Lyapounov*) *Let $\{X_n\}$ be a sequence of independent random variables such that $\mathbb{E}|X_k|^{2+\delta} < \infty$ for some $\delta > 0$ and $k = 1$, 2, ... Let*

$$\Sigma_{k=1}^n \mathbb{E}|X_k|^{2+\delta} = o(B_n^{2+\delta}) \quad as \ n \to \infty,$$

where B_n is defined in Theorem 0.54. Then condition (a) and (b) of Theorem 0.54 hold.

0.8   Branching process in a random environment

As usual, we write $\mathbb{N}^* = \{1, 2, \cdots\}$, $\mathbb{N} = \{0\}\bigcup \mathbb{N}^*$, $\mathbb{R}_+ = [0, \infty)$, and

$$\mathbb{U} = \bigcup_{n=0}^{\infty} (\mathbb{N}^*)^n$$

for the set of all finite sequences, where $(\mathbb{N}^*)^0 = \{\emptyset\}$ contains the null sequence $\emptyset$. We also write $I = (\mathbb{N}^*)^{\mathbb{N}^*}$ for the set of all infinite sequences. For $u \in \mathbb{U}$, write $|u| = n$ for the length of u, and $u|k = u_1 u_2 \cdots u_k$ ($k \leq n$) for the curtailment of u after k terms. For $|u| = n$ and $v \in \mathbb{U}$ or I, write uv for the sequence obtained by juxtaposition. We partially order $\mathbb{U}$ or I by writing $u \leq v$ to mean that for some $u' \in \mathbb{U}$, $v = uu'$.

Let $\{N_u : u \in \mathbb{U}\}$ be a family of i.i.d. random variables with values in $\mathbb{N}$, defined on some probability space $(\Omega, \mathcal{F}, \mathbb{P})$. For simplicity, we write N for $N_\emptyset$.

Let $\mathbb{T} = \mathbb{T}(\omega)$ be the usual Galton-Watson tree (1986, [107]) with defining elements $\{N_u : u \in \mathbb{U}\}$: (a) $\emptyset \in \mathbb{T}$; (b) if $ui \in \mathbb{T}$, then $u \in \mathbb{T}$; (c) if $u \in \mathbb{T}$ and $i \in \mathbb{N}^*$, then $ui \in \mathbb{T}$ if and only if $1 \leq i \leq N_u$. Here the null sequence $\emptyset$ (of length 0) represents the initial particle; ui represents the ith child of u; N_u represents the number of offspring of the particle u.

Let $\zeta = (\zeta_0, \zeta_1, \cdots)$ be a sequence of i.i.d. random variables, taking values in some space Θ, whose realization corresponds to a sequence of probability distributions on $\mathbb{N}$:

$$p(\zeta_n) = \{p_i(\zeta_n) : i \geq 0\}, \text{ where } p_i(\zeta_n) \geq 0, \sum_{i=0}^{\infty} p_i(\zeta_n) = 1.$$

A branching process $(Z_n)_{n\geq 0}$ in the random environment ζ (B.P.R.E.) is a family of time-inhomogeneous branching processes (see e.g. Athreya and Karlin (1971, [8, 9]), Athreya and Ney (1972, [10])): given the environment ζ, the process $(Z_n)_{n\geq 0}$ acts as a Galton-Watson process in varying environments with offspring distributions $p(\zeta_n)$ for particles in nth generation, $n \geq 0$. Let $\mathbb{T}_n$ be the set of all individuals of generation n, denoted by sequences u of positive integers of length $|u| = n$: as usual, the initial particle is denoted by the empty sequence $\emptyset$ (of length 0); if $u \in \mathbb{T}_n$, then $ui \in \mathbb{T}_{n+1}$ if and only if $1 \leq i \leq X_u$. By definition,

$$Z_0 = 1 \quad \text{and} \quad Z_{n+1} = \sum_{u\in \mathbb{T}_n} X_u \quad \text{for} \quad n \geq 0,$$

where conditioned on ζ, $\{X_u : |u| = n\}$ are integer-valued random variables with common distribution $p(\zeta_n)$; all the random variables X_u, indexed by finite sequences of integers u, are conditionally independent of each other. The classical Galton-Watson process corresponds to the case where all ζ_n are the same constant.

Let $(\Gamma, \mathbb{P}_\zeta)$ be the probability space under which the process is defined when the environment ζ is given. Therefore under $\mathbb{P}_\zeta$, the random variables X_u are independent of each other, and have the common law $p(\zeta_n)$ if $|u| = n$. The probability $\mathbb{P}_\zeta$ is usually called quenched law. The total probability space can be formulated as the product space $(\Gamma \times \Theta^\mathbb{N}, \mathbb{P})$, where $\mathbb{P} = \mathbb{P}_\zeta \otimes \tau$ in the sense that for all measurable and positive functions g, we have

$$\int g\,\mathrm{d}\mathbb{P} = \int \int g(\zeta, y)\mathrm{d}\mathbb{P}_\zeta(y)\mathrm{d}\tau(\zeta),$$

where τ is the law of the environment ζ. The total probability $\mathbb{P}$ is called annealed law. The quenched law $\mathbb{P}_\zeta$ may be considered to be the conditional probability of the annealed law $\mathbb{P}$ given ζ. The expectation with respect to $\mathbb{P}_\zeta$ (resp. $\mathbb{P}$) will be denoted by $\mathbb{E}_\zeta$ (resp. $\mathbb{E}$).

For $n \geq 0$, write

$$m_n(p) = m_n(p, \zeta) = \sum_{i=0}^{\infty} i^p p_i(\zeta_n) \ \text{ for } p > 0, \quad m_n = m_n(1),$$

$$P_0 = 1 \quad \text{and} \quad P_n = m_0 \cdots m_{n-1} \text{ if } n \geq 1.$$

Then $\mathbb{E}_\zeta X_u^p = m_n(p)$ if $|u| = n$, and $\mathbb{E}_\zeta Z_n = P_n$ for each n.

Let

$$m = e^\alpha.$$

It is well-known that under $\mathbb{P}_\zeta$,

$$W_n = \frac{Z_n}{P_n} \quad (n \geq 0)$$

forms a nonnegative martingale with respect to the filtration

$$\mathcal{E}_0 = \{\emptyset, \Omega\} \quad \text{and} \quad \mathcal{E}_n = \sigma\{\zeta, X_u : |u| < n\} \quad \text{for } n \geq 1.$$

It follows that $(W_n, \mathcal{E}_n)$ is also a martingale under $\mathbb{P}$. Let

$$W = \lim_{n \to \infty} W_n,$$

where the limit exists a.s. by the martingale convergence theorem, and $\mathbb{E}W \leq 1$ by Fatou's lemma.

We now consider the branching measure μ_ω in the random environment ζ, defined as follows. For the Galton-Watson tree $\mathbb{T} = \mathbb{T}(\omega)$ of branching process in the random environment ζ, we denote the boundary of $\mathbb{T}$ as

$$\partial\mathbb{T} = \{u \in I : (u|n) \in \mathbb{T} \text{ for all } n \in \mathbb{N}\}.$$

As a subset of I, $\partial \mathbb{T}$ is a metrical and compact topological space with

$$B_u = \{v \in \partial \mathbb{T} : u \leq v\} \ (u \in \mathbb{T})$$

its topological basis, and with

$$d(u, v) \doteq e^{-\min\{|u|, |v|\}}$$

a possible metric. If $u \in \mathbb{T}(\omega)$, we write $\mathbb{T}_u(\omega) = \{v \in U : uv \in \mathbb{T}(\omega)\}$ for the shifted tree of $\mathbb{T}(\omega)$ at u. Let $\mu = \mu_\omega$ be the branching measure in the random environment ζ on $\partial \mathbb{T}$: it is the unique Borel measure such that for all $u \in \mathbb{T}$,

$$\mu(B_u) = \frac{W_u}{P_{|u|}},$$

where

$$W_u = \lim_{n \to \infty} \frac{\sharp\{v \in \mathbb{T}_u : |v| = n\}}{\prod_{j=|u|}^{n+|u|-1} m_j}$$

($\sharp\{\cdot\}$ denotes the cardinality of the set $\{\cdot\}$), and

$$\mathbb{P}_\zeta(W_u \in \cdot) = \mathbb{P}_{\theta|u|\zeta}(W \in \cdot).$$

From the definition of W_u, one can easily check that

$$\mu(B_u) = W \lim_{n \to \infty} \frac{\sharp\{v \in \mathbb{T}_n : u \leq v\}}{\sharp\{v \in \mathbb{T} : |v| = n\}} \qquad \text{for all } u \in \mathbb{T}(\omega).$$

The branching measure is important in the study of branching processes. In deterministic environment case, it has been studied by many authors: see for example Joffe (1978, [63]), O'Brien (1980, [108]), Hawkes (1981, [45]), Lyons et al. (1995, [100]), Liu (1996, [87]; 2000, [90]; 2001, [93]), Liu and Rouault (1996, [95]), Liu and Shieh (1999, [97]), Shieh and Taylor (2002, [115]), Duquesne (2009, [33]), Kinnison and Mörters (2010, [68]).

For each $u \in \partial \mathbb{T}$, let $\underline{d}(\mu, u)$ and $\bar{d}(\mu, u)$ be the lower and upper local dimensions of μ at u:

$$\underline{d}(\mu, u) = \liminf_{n \to \infty} \frac{-\log \mu(B_{u|n})}{n}, \qquad \bar{d}(\mu, u) = \limsup_{n \to \infty} \frac{-\log \mu(B_{u|n})}{n}.$$

Let

$$m(n) = \min_{u \in \mathbb{T}_n} \mu(B_u) = \min_{u \in \partial \mathbb{T}} \mu(B_{u|n}) \quad \text{and} \quad M(n) = \max_{u \in \mathbb{T}_n} \mu(B_u) = \max_{u \in \partial \mathbb{T}} \mu(B_{u|n}).$$

0.9 Multiplicative cascades in a random environment

As usual, we write $\mathbb{N}^* = \{1, 2, \cdots\}$, $\mathbb{R}_+ = [0, \infty)$, $\mathbb{R} = (-\infty, \infty)$ and

$$\mathbb{U} = \bigcup_{n=0}^{\infty} (\mathbb{N}^*)^n$$

for the union of all finite sequences, where $(\mathbb{N}^*)^0 = \{\emptyset\}$ contains the null sequence $\emptyset$. We describe the model of Mandelbrot's multiplicative cascades in a random environment as follows. Let $\zeta = (\zeta_0, \zeta_1, \cdots) = (\zeta_n)_{n \geq 0}$ be a sequence of i.i.d. random variables taking values in some space Θ, so that each realization of ζ_n corresponds to a probability distribution $F_n(\zeta) = F(\zeta_n)$ on $\mathbb{R}_+$. Suppose that when the environment ζ is given, $\{W_u, u \in \mathbb{U}\}$ is a family of totally independent random variables with values in $\mathbb{R}_+$; all the random variables are defined on some probability space $(\Gamma, \mathbb{P}_\zeta)$; for $u \in \mathbb{U}$, each W_{ui} $(1 \leq i \leq r)$ has distribution $F_n(\zeta) = F(\zeta_n)$ if $|u| = n$, where $|u|$ denotes the length of u. For simplicity, we write W_i for $W_{\emptyset i}$, $1 \leq i \leq r$. The total probability space can be formulated as the product space $(\Gamma \times \Theta, \mathbb{P})$, where $\mathbb{P} = \mathbb{P}_\zeta \otimes \tau$ in the sense that for all measurable and positive functions g, we have

$$\int g \mathrm{d}\mathbb{P} = \int \int g(\zeta, y) \mathrm{d}\mathbb{P}_\zeta(y) \mathrm{d}\tau(\zeta),$$

where τ is the law of the environment ζ. The expectation with respect to $\mathbb{P}_\zeta$ (resp. $\mathbb{P}$) will be denoted by $\mathbb{E}_\zeta$ (resp. $\mathbb{E}$).

Suppose that $\mathbb{E}_\zeta W_1 = 1$ a.s. and $\mathbb{P}(W_1 = 1) < 1$.

Let $\mathcal{F}_0$ be the trivial σ-algebra, and for $n > 1$, let $\mathcal{F}_{n-1}$ be the σ-algebra generated by $\{W_{u_1}, \cdots, W_{u_1 \cdots u_{n-1}} : 1 \leq u_1, \cdots, u_{n-1} \leq r\}$. For $r = 2, 3, \cdots$, let $Z^{(r)}$ be the Mandelbrot's variable in the random environment ζ associated with W_u $(u \in \mathbb{U}/\emptyset)$ and parameter r:

$$Z^{(r)} := \lim_{n \to \infty} Y_n^{(r)},$$

where

$$Y_n^{(r)} = \sum_{1 \leq u_1, \cdots, u_n \leq r} \frac{W_{u_1} \cdots W_{u_1 \cdots u_n}}{r^n}.$$

Let $\mathbb{P}_{\theta\zeta}$ be the probability for the shifted environment $\theta\zeta$. It is easily seen that $Z = Z^{(r)}$ satisfies the following distributional equation:

$$Z^{(r)} = \frac{1}{r} \sum_{i=1}^{r} W_i Z_i^{(r)},$$

where $Z_i^{(r)}$ are non-negative random variables, which can be chosen independent of each other and independent of $\{W_i, 1 \le i \le r\}$ under $\mathbb{P}_\zeta$. Z is a non-negative random variable independent of $Z_i^{(r)}$ and independent of $\{W_i, 1 \le i \le r\}$ under $\mathbb{P}_\zeta$, $\mathbb{P}_\zeta\{Z_i^{(r)} \in \cdot\} = \mathbb{P}_{\theta\zeta}\{Z^{(r)} \in \cdot\}$. In terms of Laplace transforms $\phi_\zeta^{(r)}(t) = \mathbb{E}_\zeta \exp\{tZ^{(r)}\}$, the equation reads

$$\phi_\zeta^{(r)}(t) = \left[\mathbb{E}_\zeta \phi_{\theta\zeta}^{(r)}(tW_1/r)\right]^r, \quad \text{a.s.,} \quad t \le 0.$$

In the deterministic environment case, the model was first introduced by Mandelbrot (1974, [101]) and is referred to as "microcanonique". For one choice of W_1, $Y_n^{(r)}$ represents a stochastic model for turbulence of Yaglom (1974, [102]), and if $0 < \mathbb{P}(W_1 = 1) = 1 - \mathbb{P}(W_1 = 0)$, $r^n Y_n^{(r)}$ is the n-th generation size of a simple birth-death process. For fixed r, the properties of $Z^{(r)}$ and related subjects have been studied by many authors; see, for example, Kahane and Peyrière (1976, [64]), Durrett and Liggett (1983, [34]), Guivarc'h (1990, [38]), Holley and Waymire (1992, [48]). See also Collet and Koukiou (1992, [27]), Liu (1997, [88]; 1998, [89]; 2000, [91]), Menshikov et $al.$ (2005, [106]), Barral et $al.$ (2010, [13, 14]) for more general results and for related topics.

0.10   Directed polymers in a random environment

For notations, as usual, we write $\mathbb{N}^* = \{1, 2, \cdots\}$, $\mathbb{N} = \{0\} \bigcup \mathbb{N}^*$ and $\mathbb{R}^+ = (0, +\infty)$.

Let $d \in \mathbb{N}^*$ and $\omega = (\omega_n)_{n\in\mathbb{N}}$ be the simple random walk on the d-dimensional integer lattice $\mathbb{Z}^d$ starting at 0, defined on a probability space $(\Omega, \mathcal{F}, \mathbb{P})$. Let $\eta = (\eta(n, x))_{(n,x)\in\mathbb{N}\times\mathbb{Z}^d}$ be a sequence of real-valued, non-constant, i.i.d. random variables defined on another probability space $(E, \mathcal{E}, \mathbb{Q})$ (we use the letter E to refer to the environment). We denote by F the common distribution function of the sequence $(\eta(n, x))_{(n,x)\in\mathbb{N}\times\mathbb{Z}^d}$. The path ω represents the directed polymer and the letter η the environment sequence $(\eta(n, x))_{(n,x)\in\mathbb{N}\times\mathbb{Z}^d}$. For any $n > 0$, define the random polymer measure μ_n on the path space $(\Omega, \mathcal{F})$ by

$$\mu_n = \frac{1}{Z_n(\beta)} \exp(\beta H_n(\omega))\mathbb{P}(d\omega),$$

where $\beta > 0$ is the inverse temperature,

$$H_n(\omega) = \sum_{j=1}^n \eta(j, \omega_j) \quad \text{and} \quad Z_n(\beta) = \mathbb{P}[\exp(\beta H_n(\omega))].$$

Let $\lambda(\beta) = \ln \mathbb{Q}\left[e^{\beta\eta(0,0)}\right] \le +\infty$ be the logarithmic moment generating function of $\eta(0,0)$. Let

$$W_n(\beta) = \mathbb{P}\left[\exp\left(\beta\sum_{j=1}^{n}\eta(j,\omega_j) - n\lambda(\beta)\right)\right]$$

be the normalized partition function if $\mathbb{Q}\left[e^{\beta|\eta(0,0)|}\right] < +\infty$ for fixed $\beta > 0$ (the condition is equivalent to $\lambda(\pm\beta) < +\infty$).

This model first appeared in the physics literature (see Huse and Henley (1985, [46])) for modeling the phase boundary of the Ising model subject to random impurities, the first mathematical study was undertaken by Imbrie and Spencer (1988, [61]) and Bolthausen (1989, [19]). For recent results, see e.g.: Albeverio and Zhou (1996, [5]), Kifer (1997, [67]), Carmona and Hu (2002, [22]; 2004, [23]), Comets *et al.* (2004, [29]), Birkner (2004, [18]), Mejane (2004, [105]), Carmona *et al.* (2006, [21]), Comets and Yoshida (2006, [31]), Comets and Vargas (2006, [30]), Liu and Watbled (2009, [98, 99]), Lacoin (2010, [73]).

Chapter 1

C.C. for R.W.S. of T.A. of B-valued M.D.

The full name of the title of Chapter 1 is complete convergence for randomly weighted sums of the triangular array of Banach space valued martingale differences.

Let $(\Omega, \mathcal{F}, \mathbb{P})$ be a probability space. Let $\{X_j\}_{j\geq 1}$ be a sequence of i.i.d. real-valued random variables defined on $(\Omega, \mathcal{F}, \mathbb{P})$ with $\mathbb{E}X_1 = 0$ and set $S_n = \sum_{j=1}^{n} X_j$ for $n \geq 1$. By the weak law of large numbers, $\mathbb{P}\{|S_n| > \varepsilon n\} \to 0$ for all $\varepsilon > 0$. Hsu and Robbins (1947, [49]) introduced the notion of complete convergence which implies a.s. convergence by the Borel-Cantelli Lemma, and showed that

$$\sum_{n=1}^{\infty} \mathbb{P}\{|S_n| > \varepsilon n\} < \infty \quad \text{for all } \varepsilon > 0 \tag{1.1}$$

if $\mathbb{E}X_1^2 < \infty$; Erdös (1949, [35]) proved that the converse also holds.

Many authors have considered the complete convergence for partial sums of the sequence (array) of real-valued random variables. Spitzer (1956, [116]) showed that

$$\sum_{n=1}^{\infty} n^{-1}\mathbb{P}\{|S_n| > \varepsilon n\} < \infty \quad \text{for all } \varepsilon > 0 \tag{1.2}$$

whenever $\mathbb{E}X_1 = 0$. Katz (1963, [66]) and Baum and Katz (1965, [15]) proved that, for $p = \frac{1}{\alpha}$ and $\alpha \geq \frac{1}{2}$, or $p > \frac{1}{\alpha}$ and $\alpha > \frac{1}{2}$,

$$\sum_{n=1}^{\infty} n^{p\alpha-2}\mathbb{P}\{|S_n| > \varepsilon n^{\alpha}\} < \infty \quad \text{for all } \varepsilon > 0 \tag{1.3}$$

if and only if $\mathbb{E}|X_1|^p < \infty$. Lai (1974, [75]) studied the limiting case of the theorem of Baum and Katz (1965, [15]) when $p > 2$ and $\alpha = \frac{1}{2}$. Gafurov and Slastnikov (1987, [36]) considered the case where $(n^{p\alpha-2})$ and (n^α) are replaced by more general sequences. Bai and Su (1985, [11]) showed that, for $p = \frac{1}{\alpha}$ and $\alpha > \frac{1}{2}$, or $p > \frac{1}{\alpha}$ and $\alpha > \frac{1}{2}$,

$$\sum_{n=1}^{\infty} n^{p\alpha-2}\ell(n)\mathbb{P}\{|S_n| > \varepsilon n^\alpha\} < \infty \quad \text{for all } \varepsilon > 0 \tag{1.4}$$

if and only if $\mathbb{E}[|X_1|^p\ell(|X_1|^{1/\alpha})] < \infty$, where $\ell(\cdot) > 0$ is a function slowly varying at $+\infty$ and monotone increasing for $p = \frac{1}{\alpha}$ and $\alpha > \frac{1}{2}$. Chow (1973, [25]) proved that, for $p \geq \frac{1}{\alpha}$ and $\alpha > \frac{1}{2}$,

$$\sum_{n=1}^{\infty} n^{p\alpha-2}\mathbb{P}\{\max_{1 \leq i \leq n} |S_i| > \varepsilon n^\alpha\} < \infty \quad \text{for all } \varepsilon > 0 \tag{1.5}$$

if and only if $\mathbb{E}|X_1|^p < \infty$. Some authors have considered the generalizations of the theorem of Baum and Katz (1965, [15]) to arrays of independent but not necessarily identically distributed real-valued random variables, see e.g. Li *et al.* (1995, [79]), Hu *et al.* (1998, [57]; 2000, [58]; 2003, [54]), Kuczmaszewska (2004, [71]), Sung *et al.* (2005, [124]), Kruglov *et al.* (2006, [70]). Let $\{(X_j, \mathcal{F}_j)\}_{j \geq 1}$ be a sequence of real-valued martingale differences defined on $(\Omega, \mathcal{F}, \mathbb{P})$, adapted to a filtration $\{\mathcal{F}_j\}_{j \geq 0}$, with $\mathcal{F}_0 = \{\emptyset, \Omega\}$. Some authors have considered the generalizations of the theorem of Baum and Katz to sequences (arrays) of real-valued martingale differences. Lesigne and Volný (2001, [78]) proved that for $p \geq 2$, $\sup_{j \geq 1} \mathbb{E}|X_j|^p < \infty$ implies

$$\mathbb{P}(|S_n| > \varepsilon n) = O(n^{-p/2}) \tag{1.6}$$

(as usual we write $a_n = O(b_n)$ if the sequence (a_n/b_n) is bounded), and that the exponent $(p/2)$ is the best possible, even for strictly stationary and ergodic sequences of real-valued martingale differences. Therefore the theorem of Baum and Katz does not hold for the sequences of real-valued martingale differences without additional conditions. Alsmeyer (1990, [6]) proved that the theorem of Baum and Katz for $p > \frac{1}{\alpha}$ and $\frac{1}{2} < \alpha \leq 1$ still holds for the sequence of real-valued martingale differences $\{(X_j, \mathcal{F}_j)\}_{j \geq 1}$ if for some $\gamma \in (\frac{1}{\alpha}, 2]$ and $q \in [1, \infty]$ with $q > (p\alpha - 1)/(\gamma\alpha - 1)$,

$$\sup_{n \geq 1} \left\| \frac{1}{n} \sum_{j=1}^{n} \mathbb{E}[|X_j|^\gamma | \mathcal{F}_{j-1}] \right\|_q < \infty, \tag{1.7}$$

where $\|\cdot\|_q$ denotes the $\mathbb{L}^q$ norm. Under a stochastically dominated condition, Wang and Hu (2014, [141]) investigated the complete convergence for the maximal partial sums of the sequence of real-valued martingale differences and got the Baum-Katz type theorems for sequences of real-valued martingale differences. Under a simple moment condition, Hao and Liu (2012, [43]; 2014, [44]) considered the complete convergence for partial sums of the array of real-valued martingale differences and extended the theorem of Baum and Katz to arrays of real-valued martingale differences.

Some authors have also considered the complete convergence for partial sums of the sequence of Banach space valued random elements. Let $(\mathbb{B}, \|\cdot\|)$ be a real separable Banach space. Let $\{X_j\}_{j\geq1}$ be a sequence of i.i.d. $\mathbb{B}$-valued random elements defined on $(\Omega, \mathcal{F}, \mathbb{P})$ and set $S_n = \sum_{j=1}^{n} X_j$ for $n \geq 1$. Jain (1975, [62]) proved that

$$\sum_{n=1}^{\infty} n^{-1}\mathbb{P}\{\|S_n\| > \varepsilon n\} < \infty \quad \text{for all } \varepsilon > 0 \tag{1.8}$$

if and only if $\mathbb{E}\|X_1\| < \infty$ and $\mathbb{E}X_1 = 0$. Yang and Wang (1986, [148]) proved that, for $\alpha > \frac{1}{2}$,

$$\sum_{n=1}^{\infty} n^{-1}\mathbb{P}\{\|S_n\| > \varepsilon n^{\alpha}\} < \infty \quad \text{for all } \varepsilon > 0 \tag{1.9}$$

if and only if $\mathbb{E}\|X_1\|^{1/\alpha} < \infty$ and $S_n/n^{\alpha} \to 0$ in probability. Suppose that $\left\{S_j/j^{\frac{1}{2\wedge p}}\right\}_{j\geq1}$ is bounded in probability with $p > 0$ (given $\varepsilon > 0$, there exists $A > 0$ such that for all $j \geq 1$, $\mathbb{P}\{\|S_j/j^{\frac{1}{2\wedge p}}\| > A\} < \varepsilon$). Jain (1975, [62]) proved that, for $p > \frac{1}{\alpha}$ and $\alpha > \frac{1}{2}$,

$$\sum_{n=1}^{\infty} n^{p\alpha-2}\mathbb{P}\{\|S_n\| > \varepsilon n^{\alpha}\} < \infty \quad \text{for all } \varepsilon > 0 \tag{1.10}$$

if and only if $\mathbb{E}\|X_1\|^p < \infty$. Let $\{(X_j, \mathcal{F}_j)\}_{j\geq1}$ be a sequence of $\mathbb{B}$-valued martingale differences defined on $(\Omega, \mathcal{F}, \mathbb{P})$, adapted to a filtration $\{\mathcal{F}_j\}_{j\geq0}$, with $\mathcal{F}_0 = \{\emptyset, \Omega\}$. Assume that $\{X_j\} \prec X$ for some positive random variable X in the meaning that $Q_X \geq \sup_{j\geq1} Q_{\|X_j\|}$, where the quantile function $Q_X(u) = L_X^{-1}(u) = \inf\{t \geq 0 : L_X(t) \leq u\}$ is the generalized inverse of the upper tail function $L_X(t) = \mathbb{P}\{X > t\}$. Dedecker and Merlevède (2008, [32]) proved that, for $1 < p < 2$ and $1/p \leq \alpha \leq 1$,

$$\sum_{n=1}^{\infty} n^{p\alpha-2}\mathbb{P}\left\{\max_{1\leq i\leq n} \|S_i\| \geq \varepsilon n^{\alpha}\right\} < \infty \quad \text{for all } \varepsilon > 0 \tag{1.11}$$

if $\mathbb{E}X^p < \infty$ and $\mathbb{B}$ is r-smooth for some $r > p$. Dedecker and Merlevède (2008, [32]) also proved that

$$\sum_{n=1}^{\infty} n^{-1}\mathbb{P}\left\{\max_{1\leq i\leq n} \|S_i\| \geq \varepsilon n\right\} < \infty \quad \text{for all } \varepsilon > 0 \tag{1.12}$$

if $\mathbb{E}(X \ln^+ X) < +\infty$ and $\mathbb{B}$ is super-reflexive. Let $\{a_j\}_{j \geq 1}$ be a sequence of positive constants such that $a_n \uparrow \infty$ and $1 < \liminf\limits_{n \to \infty} \frac{a_{2n}}{a_n} \leq \limsup\limits_{n \to \infty} \frac{a_{2n}}{a_n} < \infty$. Li *et al.* (2007, [82]) proved that

$$\sum_{n=1}^{\infty} n^{-1} \mathbb{P}\{\|S_n\| \geq \varepsilon a_n\} < \infty \quad \text{for all } \varepsilon > 0 \tag{1.13}$$

if and only if $\lim\limits_{n \to \infty} \frac{S_n}{a_n} = 0$ a.s.

It is well known that partial sum is a particular case of weighted sum. Many linear models in statistics based on a random sample involve weighted sums of dependent random variables. Examples include least-squares estimators, nonparametric regression function estimators, and jackknife estimates, among others. In this respect, the study of convergence for these weighted sums has impact on statistics, probability, and their applications. (cf. [133]) In fact, the study of convergence for these weighted sums also has impact on systems theory, mathematical physics, applied economics, and their applications. Thus the complete convergence for the weighted sums seems more important. (cf. [143])

Many authors have considered the complete convergence for weighted sums of the sequence (array) of real-valued random variables. Gut (1993, [39]) provided some necessary and sufficient conditions for complete convergence of the Cesàro means of the sequence of i.i.d. random variables, Lanzinger and Stadtmüller (2003, [76]) considered some necessary and sufficient conditions for complete convergence of weighted sums of the sequence of i.i.d. random variables. Li *et al.* (1995, [79]) obtained some sufficient conditions for complete convergence of weighted sums of the sequence of independent random variables, Wang *et al.* (1998, [143]) proved some necessary and sufficient conditions for complete convergence of weighted sums of the sequence of independent random variables. Some authors have considered the complete convergence for weighted sums of the sequence (array) of real-valued martingale differences: Stout (1968, [118]) obtained a sufficient condition for complete convergence of weighted sums of the sequence of martingale differences, Yu (1990, [150]) also showed a sufficient condition for complete convergence of weighted sums of the sequence of martingale differences (see also the references therein), Wang *et al.* (2012, [142]) provided some sufficient conditions for complete convergence of weighted sums of the sequence of martingale differences, under a simple moment condition, Hao and Liu (2012, [43]; 2014, [44]) proved a necessary and sufficient condition for complete convergence of specially single-indexed weighted sums of the sequence of identically distributed martingale differences, Yang *et al.* (2013, [147]) gave a sufficient condition for complete convergence of randomly weighted sums of the sequence of martingale differences, Ghosal and Chandra (1998, [37]) proved some sufficient conditions for complete convergence of weighted sums of the array of martingale differences. Few authors have considered the complete convergence for double-indexed randomly weighted sums of mixing random variables. (cf. [138, 134])

Some authors have also considered the complete convergence for weighted sums of the sequence (array) of Banach space valued random elements. Tómács (2005, [135]) obtained a sufficient condition for complete convergence of weighted sums of the triangular array of rowwise independent random elements, Li et al. (1995, [79]), Hernández et al. (2007, [47]) gave some sufficient conditions for complete convergence of weighted sums of the array of rowwise independent random elements; under a stochastically dominated condition, Hu et al. (1989, [53]), Wang et al. (1993, [140]), Sung (1997, [119]), Hu et al. (1999, [56]) proved some sufficient conditions for complete convergence of weighted sums of the triangular array of rowwise independent random elements, Hu et al. (1999, [56]), Ahmed et al. (2002, [4]), Hu et al. (2003, [52]), Chen et al. (2006, [24]), Sung and Volodin (2006, [122]; 2011, [123]), Sung (2010, [121]), Qiu et al. (2012, [112]) provided some sufficient conditions for complete convergence of weighted sums of the array of rowwise independent random elements, Qiu et al. (2012, [112]) showed a sufficient condition for complete convergence of weighted sums of the array of any random elements, Thanh and Yin (2011, [133]) considered a sufficient condition for complete convergence of randomly weighted sums of the triangular array of rowwise independent random elements and a sufficient condition for complete convergence of randomly weighted sums of the triangular array of any random elements; Dedecker and Merlevède (2008, [32]) proved a sufficient condition for complete convergence of weighted sums of the weakly dependent sequence of random elements; under a simple moment condition, Hao (2013, [42]) obtained a necessary and sufficient condition for complete convergence of specially single-indexed weighted sums of the sequence of identically distributed martingale differences, a sufficient condition for complete convergence of weighted sums of the triangular array of identically distributed martingale differences and two sufficient conditions for complete convergence of weighted sums of the array of martingale differences. Starting from the 1970s, the random nature of many problems arising in the applied sciences is noted. This leads to mathematical models which deal with the limiting behaviour of weighted sums of random elements in normed linear spaces, where the weights are random variables. Taylor (1972, [127]) and Taylor and Padgett (1974, [129]; 1976, [130]) obtained some basic results by considering random weights $(A_j)_{j\geq1}$. From 1978 on, it begins to be studied directly the convergence of randomly weighted sums of random elements in separable Banach spaces or in separable general normed linear spaces, such as Wei and Taylor (1978, [145, 146]), Taylor and Calhoun (1983, [128]), Taylor et al. (1984, [131]), Ordóñez Cabrera (1988, [109]), Adler et al. (1992, [2]), Wang and Rao (1995, [139]) and Hu and Chang (1999, [51]). The limiting behaviour of randomly weighted sums plays an important role in various applied and theoretical problems. One can see the example of Rosalsky and Sreehari (1998, [114]), in queueing theory. (cf. [55])

As information, we mention that some authors have also considered the complete convergence for weighted sums of the sequence (array) of Banach space valued random elements in p-type Banch space setting. Liang (2000, [84]), Liang and Wang (2001, [85]) obtained some necessary and sufficient conditions for complete convergence of weighted sums of the sequence of dependent Banach space valued random elements in Banach space of type p. Hernández et al. (2007, [47]) gave a sufficient condition for complete convergence of weighted sums of the array of rowwise independent Banach space valued random elements in Banach space of type p.

For one thing, most literature in Banach space setting is about the weighted sums of the (triangular) array of rowwise independent Banach space valued random elements, few literature (cf. [42]) is about the weighted sums of the (triangular) array of Banach space valued martingale differences. For another thing, some results of Hao (2013, [42]) required that the martingale differences are identically distributed, which is too strong to apply conveniently. Further more, a number of researchers considered the weighted sums of Banach space valued random elements where the weights are real constants. To the best of our knowledge, there has only been few results (cf. [133]) on the weighted sums of Banach space valued random elements in which the weights are random variables. Thus we consider the convergence rates for double-indexed randomly weighted sums of the triangular array of Banach space valued martingale differences. The new challenge is that one has to treat a weighted sum of the triangular array of Banach space valued martingale differences whose weights are random with double indices.

Our main objective is twofold. One is to establish the complete convergence for double-indexed randomly weighted sums of the triangular array of Banach space valued martingale differences. The other is to extend Theorem 2 of Dedecker and Merlevède (2008, [32]) from partial sums of the sequence of Banach space valued martingale differences to double-indexed randomly weighted ones of the triangular array of Banach space valued martingale differences.

The rest of the chapter is organized as follows. In Section 1.1 we consider the convergence rates for double-indexed randomly weighted sums of the triangular array of Banach space valued martingale differences, and extend Theorem 2 of Dedecker and Merlevède (2008, [32]). Section 1.2 is devoted to some preliminary facts which are used in Section 1.3, where the proof of the main result is given. ∎

1.1   Main result

In this section, we consider the convergence rates in the law of large numbers for double-indexed randomly weighted sums of the triangular array of Banach space valued martingale differences.

Let $(\Omega, \mathcal{F}, \mathbb{P})$ be a probability space and $(\mathbb{B}, \|\cdot\|)$ be a separable Banach space. Following Pisier (1975, [110]) and Dedecker and Merlevède (2008, [32]), we say that a Banach space $(\mathbb{B}, \|\cdot\|)$ is r-smooth $(1 < r \leq 2)$ if there exists an equivalent norm $\|\cdot\|_{\mathbb{B}}$ such that

$$\sup_{t>0}\left\{\frac{1}{t^r}\sup\{\|x + ty\|_{\mathbb{B}} + \|x - ty\|_{\mathbb{B}} - 2 : \|x\|_{\mathbb{B}} = \|y\|_{\mathbb{B}} = 1\}\right\} < \infty.$$

A Banach space is said to be super-reflexive if it is r-smooth for some $1 < r \leq 2$. By Proposition 2 of Assouad (1975, [7]), we know that if $\mathbb{B}$ is r-smooth and separable, then there exists a constant D such that, for any sequence of $\mathbb{B}$-valued martingale differences $(X_j)_{j\geq 1}$,

$$\mathbb{E}\left\|\sum_{j=1}^{n}X_j\right\|^r \leq D\sum_{j=1}^{n}\mathbb{E}\|X_j\|^r. \tag{1.14}$$

For every $n \geq 1$, let $\mathcal{F}_{n0} = \{\emptyset, \Omega\} \subset \mathcal{F}_{n1} \subset \cdots \subset \mathcal{F}_{nn}$ be an increasing sequence of sub-σ-fields of $\mathcal{F}$. For each $n \geq 1$, let $\{(X_{nj}, \mathcal{F}_{nj})\}_{j=1}^{n}$ be a sequence of $\mathbb{B}$-valued martingale differences defined on $(\Omega, \mathcal{F}, \mathbb{P})$, adapted to the filtration $(\mathcal{F}_{nj})$: that is, for every $1 \leq j \leq n$ and every $n \geq 1$, X_{nj} is $\mathcal{F}_{nj}$-measurable and belongs to $\mathbb{L}_{\mathbb{B}}^1$, and $\mathbb{E}[X_{nj}|\mathcal{F}_{n,j-1}] = 0$ a.s. Set for $n \geq 1$,

$$S_{n0} = 0, \ S_{nk} = \sum_{j=1}^{k}X_{nj}, \ 1 \leq k \leq n.$$

We shall call the double sequence $\{(X_{nj}, \mathcal{F}_{nj}), 1 \leq j \leq n, n \geq 1\}$ a triangular array of $\mathbb{B}$-valued martingale differences.

Definition 1.1 *For any non-increasing cadlag function f from $\mathbb{R}^+$ to $\mathbb{R}^+$, define the generalized inverse $f^{-1}(u) = \inf\{t \geq 0 : f(t) \leq u\}$. For any $\mathbb{B}$-valued random element X, define the upper tail function $L_{\|X\|}(t) = \mathbb{P}(\|X\| > t)$ and the quantile function $Q_{\|X\|} = L_{\|X\|}^{-1}$. Let $\{X_{nj}, 1 \leq j \leq n, n \geq 1\}$ be a triangular array of $\mathbb{B}$-valued random elements. Following Dedecker and Merlevède (2008, [32]), we write $(X_{nj}) \prec X$ if there exists a $\mathbb{B}$-valued random element X such that $Q_{\|X\|} \geq \sup_{n,j:1\leq j\leq n}Q_{\|X_{nj}\|}$.*

Let $\mathcal{F}_0 = \{\emptyset, \Omega\} \subset \mathcal{F}_1 \subset \cdots$ be an increasing sequence of sub-σ-fields of $\mathcal{F}$ and let $\{(X_j, \mathcal{F}_j)\}_{j\geq 1}$ be a sequence of $\mathbb{B}$-valued martingale differences. Assume that $(X_j) \prec X$ for some positive random variable X in the meaning that $Q_X \geq \sup_{j\geq 1}Q_{\|X_j\|}$. Let $1 < p < 2$. Dedecker and Merlevède (2008, [32]) proved that, for any $1 \leq 1/\alpha \leq p$, (1.11) holds for all $\varepsilon > 0$ if $\mathbb{E}X^p < \infty$ and $\mathbb{B}$ is r-smooth for some $r > p$. Dedecker and Merlevède (2008, [32]) also proved that (1.12) holds for all $\varepsilon > 0$ if $\mathbb{E}(X\ln^+ X) < \infty$ and $\mathbb{B}$ is super-reflexive. This is a nice result, nevertheless the conditions are not always satisfied in applications, for example: in applications instead of a $\mathbb{B}$-valued single martingale we often need to consider a $\mathbb{B}$-valued martingale array: for example when we use the decomposition of a random sequence (S_n) into B-valued martingale differences (such as in the study of directed polymers in a random environment), the summands usually depend on n: $S_n = \sum_{j=1}^{n} X_{nj}$, $X_{nj} = \mathbb{E}[S_n|\mathcal{F}_j] - \mathbb{E}[S_n|\mathcal{F}_{j-1}]$, where $\mathcal{F}_0 = \{\emptyset, \Omega\}$ and $\mathcal{F}_i = \sigma(S_1, \cdots, S_i)$ for $i \geq 1$. In the following we give a sufficient condition for the complete convergence of double-indexed randomly weighted sums of the triangular array $\{(X_{nj}, \mathcal{F}_{nj}), 1 \leq j \leq n, n \geq 1\}$ of B-valued martingale differences and extend Theorem 2 of Dedecker and Merlevède (2008, [32]).

Let $\ell(\cdot) > 0$ be a function slowly varying at ∞. Recall that a function $\ell(x) > 0$ slowly varying at ∞ has the representation form

$$\ell(x) = c(x) \exp\left\{ \int_{x_0}^{x} \frac{\varepsilon(u)}{u} du \right\} \quad (x \geq x_0) \tag{1.15}$$

for some $x_0 > 0$, where $c(\cdot)$ is measurable and $c(x) \to c \in (0, \infty)$, $\varepsilon(x) \to 0$ as $x \to \infty$. The function $c(\cdot)$ plays no role for our purpose. We can choose $c(x) \equiv 1$ without loss of generality.

Theorem 1.1 *Suppose that $(X_{nj}) \prec X$ for some B-valued random element X. Let $\{A_{nj}, 1 \leq j \leq n, n \geq 1\}$ be a triangular array of real-valued random variables. Assume that the random variable A_{nj} is independent of $\mathcal{F}_{nj}$ for all $n \geq 1$ and $1 \leq j \leq n$.*
(a) Let $1/2 < \alpha \leq 1$, $1 < p < 2$ and $\ell(\cdot) > 0$ be a slowly varying function at ∞. If $\mathbb{E}\left[\|X\|^p \ell(\|X\|^{1/\alpha})\right] < \infty$, B is r-smooth for some $r > p$ and

$$\sum_{j=1}^{n} \mathbb{E}|A_{nj}|^r = O(n), \tag{1.16}$$

then

$$\sum_{n=1}^{\infty} n^{p\alpha-2}\ell(n)\mathbb{P}\left\{ \sup_{1 \leq k \leq n} \left\| \sum_{j=1}^{k} A_{nj}X_{nj} \right\| \geq \varepsilon n^\alpha \right\} < \infty \quad \text{for all } \varepsilon > 0. \tag{1.17}$$

(b) If $\mathbb{E}\left[\|X\| \ln^+ \|X\|\right] < \infty$, B is super-reflexive and

$$\sum_{j=1}^{n} \mathbb{E}|A_{nj}| - O(n), \tag{1.18}$$

then

$$\sum_{n=1}^{\infty} n^{-1}\mathbb{P}\left\{ \sup_{1 \leq k \leq n} \left\| \sum_{j=1}^{k} A_{nj}X_{nj} \right\| \geq \varepsilon n \right\} < \infty \quad \text{for all } \varepsilon > 0. \tag{1.19}$$

For a sequence of single $\mathbb{B}$-valued martingale differences, in the case where $A_{nj} = 1$ for all $n \geq 1$ and $1 \leq j \leq n$, X is a positive random variable and $\ell(\cdot) = 1$, by Theorem 1.1, we get Theorem 2 of Dedecker and Merlevède (2008, [32]).

Corollary 1.1 *Let* $(X_j)_{j \geq 1}$ *be a sequence of i.i.d. real-valued random variables with* $\mathbb{E}X_i = 0$.
(a) Let $1/2 < \alpha \leq 1$ *and* $1 < p < 2$. *If* $\mathbb{E}|X_1|^p < \infty$, *then*

$$\sum_{n=1}^{\infty} n^{p\alpha - 2} \mathbb{P} \left\{ \sup_{1 \leq k \leq n} \left| \sum_{j=1}^{k} X_j \right| \geq \varepsilon n^{\alpha} \right\} < \infty \quad \text{for all } \varepsilon > 0. \qquad (1.20)$$

(b) If $\mathbb{E}\left[|X_1| \ln^+ |X_1| \right] < \infty$, *then*

$$\sum_{n=1}^{\infty} n^{-1} \mathbb{P} \left\{ \sup_{1 \leq k \leq n} \left| \sum_{j=1}^{k} X_j \right| \geq \varepsilon n \right\} < \infty \quad \text{for all } \varepsilon > 0. \qquad (1.21)$$

Notice that we consider the supremum of partial sums in Corollary 1.1.   ∎

1.2   Preliminaries

The proof of Theorem 1.1 will need the following two lemmas.

Lemma 1.1 (cf. Hao [42], Lemma 23) *Let* $\alpha > -1$. *Then for some* $n_0, c_1, c_2 > 0$ *and all* $N \geq n_0$,

$$c_1 N^{\alpha + 1} \ell(N) \leq \sum_{n=1}^{N} n^{\alpha} \ell(n) \leq c_2 N^{\alpha + 1} \ell(N).$$

Proof. Without loss of generality, we suppose that $\ell(s)$ has the form (1.15) with $c(s) \equiv 1$. Therefore, for $\delta \in (0, \alpha + 1)$, $s^{\delta} \ell(s)$ is increasing in $[n_1, \infty)$ for some $n_1 > 0$ large enough. Consequently, for some positive constants c_0, c_2, c_3 (which may depend on n_1) and all $N \geq n_1$,

$$\sum_{n=1}^{N} n^{\alpha} \ell(n) = \sum_{n=1}^{N} n^{\alpha-\delta} n^{\delta} \ell(n)$$

$$\leq c_0 + \sum_{n=n_1}^{N} n^{\alpha-\delta} n^{\delta} \ell(n)$$

$$\leq c_0 + N^{\alpha+1} \ell(N) \sum_{n=1}^{N} (\frac{n}{N})^{\alpha-\delta} \frac{1}{N}$$

$$\leq c_0 + N^{\alpha+1} \ell(N) c_3 \quad \left(\text{as } \lim_{N \to \infty} \sum_{n=1}^{N} (\frac{n}{N})^{\alpha-\delta} \frac{1}{N} = \int_0^1 x^{\alpha-\delta} dx < \infty \right)$$

$$\leq c_2 N^{\alpha+1} \ell(N).$$

Similarly, for $N \geq 2n_1$,

$$\sum_{n=1}^{N} n^{\alpha} \ell(n) \geq \sum_{N/2 \leq n \leq N} n^{\alpha-\delta} n^{\delta} \ell(n)$$

$$\geq (\frac{N}{2})^{\delta} \ell(\frac{N}{2}) \sum_{N/2 < n \leq N} n^{\alpha-\delta}.$$

Since ℓ is slowly varying at ∞, by Potter's Theorem, for $A = 2$ and $\delta_1 = 1$, there exists n_2 such that for all $N \geq n_2$, $\ell(\frac{N}{2}) \geq 2^{-2} \ell(N)$. If $\delta \leq \alpha$, then $\sum_{N/2 < n \leq N} n^{\alpha-\delta} \geq (\frac{N}{2})^{\alpha-\delta} (N - [N/2]) \geq c_4 N^{\alpha+1-\delta}$ for some positive constant c_4; if $\delta > \alpha$, then $\sum_{N/2 < n \leq N} n^{\alpha-\delta} \geq N^{\alpha-\delta} (N - [N/2]) \geq c_5 N^{\alpha+1-\delta}$ for some positive constant c_5. So for some constants $c_1 > 0$, $n_0 = \max\{2n_1, n_2\}$ and all $N \geq n_0$,

$$\sum_{n=1}^{N} n^{\alpha} \ell(n) \geq c_1 N^{\alpha+1} \ell(N).$$

∎

Lemma 1.2 *Let $\alpha < -1$. Then for some n_3, c_4, $c_5 > 0$ and all $N \geq n_3$,*

$$c_4 N^{\alpha+1} \ell(N) \leq \sum_{n=N}^{\infty} n^{\alpha} \ell(n) \leq c_5 N^{\alpha+1} \ell(N).$$

Proof. Without loss of generality, we suppose that $\ell(s)$ has the form (1.15) with $c(s) \equiv 1$. Therefore, for $\delta \in (0, -\alpha - 1)$, $s^{-\delta} \ell(s)$ is decreasing in $[n_4, \infty)$ for some

$n_4 > 0$ large enough. Consequently, for some positive constant c_5 (which may depend on n_4) and all $N \geq n_4$,

$$\sum_{n=N}^{\infty} n^{\alpha} \ell(n) = \sum_{n=N}^{\infty} n^{\alpha+\delta} n^{-\delta} \ell(n)$$

$$\leq N^{\alpha+1} \ell(N) \sum_{n=N}^{\infty} \left(\frac{n}{N}\right)^{\alpha+\delta} \frac{1}{N}$$

$$\leq c_5 N^{\alpha+1} \ell(N)$$

because

$$\sum_{n=N}^{\infty} \left(\frac{n}{N}\right)^{\alpha+\delta} \frac{1}{N} \leq \sum_{n=N}^{\infty} \int_{n-1}^{n} \left(\frac{x}{N}\right)^{\alpha+\delta} d\left(\frac{x}{N}\right)$$

$$= \int_{N-1}^{+\infty} \left(\frac{x}{N}\right)^{\alpha+\delta} d\left(\frac{x}{N}\right)$$

$$= \frac{\left(\frac{N-1}{N}\right)^{\alpha+\delta+1}}{-\alpha-\delta-1} \to \frac{-1}{\alpha+\delta+1} \quad (N \to \infty).$$

Similarly, for $N \geq n_4$,

$$\sum_{n=N}^{\infty} n^{\alpha} \ell(n) \geq \sum_{n=N}^{2N} n^{\alpha+\delta} n^{-\delta} \ell(n)$$

$$\geq (2N)^{-\delta} \ell(2N) \sum_{n=N}^{2N} n^{\alpha+\delta}.$$

Since ℓ is slowly varying at ∞, by Potter's Theorem (Theorem 1.5.6 in [17, p. 25]), for $A = 2$ and $\delta_1 = 1$, there exists n_5 such that for all $N \geq n_5$, $\ell(2N) \geq 2^{-2}\ell(N)$. And we have that $\sum_{n=N}^{2N} n^{\alpha+\delta} \geq (2N)^{\alpha+\delta}(N+1) \geq c_9 N^{\alpha+1+\delta}$ for some positive constant c_9. So for some constants $c_4 > 0$, $n_3 = \max\{n_4, n_5\}$ and all $N \geq n_3$,

$$\sum_{n=N}^{\infty} n^{\alpha} \ell(n) \geq c_4 N^{\alpha+1} \ell(N).$$

∎

1.3   Proof of the main result

We get Theorem 1.1 by a refinement of the method of Dedecker and Merlevède (2008, [32]).

Proof of Theorem 1.1. (a) Note that (1.17) is automaic if $p\alpha < 1$. So, we assume that $p\alpha \geq 1$. For $n \geq 1$ and $1 \leq j \leq n$, we define the four random elements

$$X'_{nj} = X_{nj}\mathbf{1}_{\{\|X_{nj}\|\leq n^{\alpha}\}},$$
$$X''_{nj} = X_{nj}\mathbf{1}_{\{\|X_{nj}\|>n^{\alpha}\}},$$
$$Y'_{nj} = X'_{nj} - \mathbb{E}[X'_{nj}|\mathcal{F}_{n,j-1}]$$

and

$$Y''_{nj} = X''_{nj} - \mathbb{E}[X''_{nj}|\mathcal{F}_{n,j-1}].$$

Since $\{(X_{nj}, \mathcal{F}_{nj}), 1 \leq j \leq n, n \geq 1\}$ is a triangular array of $\mathbb{B}$-valued martingale differences, we have that $X_{nj} = Y'_{nj} + Y''_{nj}$. Therefore, for every $\varepsilon > 0$,

$$\mathbb{P}\left\{\sup_{1\leq k\leq n}\left\|\sum_{j=1}^{k}A_{nj}X_{nj}\right\| \geq 2\varepsilon n^{\alpha}\right\} \leq \mathbb{P}\left\{\sup_{1\leq k\leq n}\left\|\sum_{j=1}^{k}A_{nj}Y'_{nj}\right\| \geq \varepsilon n^{\alpha}\right\}$$

$$+ \mathbb{P}\left\{\sup_{1\leq k\leq n}\left\|\sum_{j=1}^{k}A_{nj}Y''_{nj}\right\| \geq \varepsilon n^{\alpha}\right\}. \quad (1.22)$$

By Markov's inequality, the independence of A_{nj} and X_{nj} for any $n \geq 1$ and $1 \leq j \leq n$, and the definition of Y''_{nj} and X''_{nj}, we obtain that

$$\mathbb{P}\left\{\sup_{1\leq k\leq n}\left\|\sum_{j=1}^{k}A_{nj}Y''_{nj}\right\| \geq \varepsilon n^{\alpha}\right\} \leq \frac{1}{\varepsilon n^{\alpha}}\mathbb{E}\sup_{1\leq k\leq n}\left\|\sum_{j=1}^{k}A_{nj}Y''_{nj}\right\|$$

$$\leq \frac{1}{\varepsilon n^{\alpha}}\sum_{j=1}^{n}\mathbb{E}\|A_{nj}Y''_{nj}\|$$

$$= \frac{1}{\varepsilon n^{\alpha}}\sum_{j=1}^{n}\mathbb{E}|A_{nj}|\mathbb{E}\|Y''_{nj}\|$$

$$\leq \frac{2}{\varepsilon n^{\alpha}}\sum_{j=1}^{n}\mathbb{E}|A_{nj}|\mathbb{E}\|X''_{nj}\|$$

$$= \frac{2}{\varepsilon n^{\alpha}}\sum_{j=1}^{n}\mathbb{E}|A_{nj}|\mathbb{E}\left(\|X_{nj}\|\mathbf{1}_{\{\|X_{nj}\|>n^{\alpha}\}}\right).$$

$$(1.23)$$

Noting that for any $n \geq 1$ and $1 \leq j \leq n$,

$$
\mathbb{E}\left(\|X_{nj}\|\mathbf{1}_{\{\|X_{nj}\|>n^\alpha\}}\right) = \int_0^1 Q_{\|X_{nj}\|\mathbf{1}_{\{\|X_{nj}\|>n^\alpha\}}}(u)du
$$

$$
\leq \int_0^1 Q_{\|X\|\mathbf{1}_{\{\|X\|>n^\alpha\}}}(u)du. \tag{1.24}
$$

For any $A > 0$, we have

$$
Q_{\|X\|\mathbf{1}_{\{\|X\|>A\}}}(u) = Q_{\|X\|}(u)\mathbf{1}_{\{u<L_{\|X\|}(A)\}}. \tag{1.25}
$$

By (1.23), (1.24) and (1.25), we obtain that

$$
\mathbb{P}\left\{\sup_{1\leq k\leq n}\left\|\sum_{j=1}^k A_{nj}Y''_{nj}\right\| \geq \varepsilon n^\alpha\right\}
$$

$$
\leq \frac{2}{\varepsilon n^\alpha}\left(\sum_{j=1}^n \mathbb{E}|A_{nj}|\right)\int_0^1 Q_{\|X\|}(u)\mathbf{1}_{\{u<L_{\|X\|}(n^\alpha)\}}du. \tag{1.26}
$$

Hence, by (1.16), we get that

$$
\mathbb{P}\left\{\sup_{1\leq k\leq n}\left\|\sum_{j=1}^k A_{nj}Y''_{nj}\right\| \geq \varepsilon n^\alpha\right\} \leq \frac{2}{\varepsilon}O(n^{1-\alpha})\int_0^1 Q_{\|X\|}(u)\mathbf{1}_{\{u<L_{\|X\|}(n^\alpha)\}}du. \tag{1.27}
$$

Since $u < L_{\|X\|}(n^\alpha)$ if and only if $n < Q_{\|X\|}^{1/\alpha}(u)$, by Lemma 1.1, we have that there exists a finite constant C depending only on p, α and ε such that

$$
\sum_{n=1}^\infty n^{\alpha p-2}\ell(n)\mathbb{P}\left\{\sup_{1\leq k\leq n}\left\|\sum_{j=1}^k A_{nj}Y''_{nj}\right\| \geq \varepsilon n^\alpha\right\}
$$

$$
\leq \frac{2}{\varepsilon}\sum_{n=1}^\infty O(n^{\alpha(p-1)-1})\ell(n)\int_0^1 Q_{\|X\|}(u)\mathbf{1}_{\{u<L_{\|X\|}(n^\alpha)\}}du
$$

$$
= \frac{2}{\varepsilon}\int_0^1 Q_{\|X\|}(u)\sum_{n=1}^\infty O(n^{\alpha(p-1)-1})\ell(n)\mathbf{1}_{\{u<L_{\|X\|}(n^\alpha)\}}du
$$

$$
= \frac{2}{\varepsilon}\int_0^1 Q_{\|X\|}(u)\sum_{n=1}^{[Q_{\|X\|}^{1/\alpha}(u)]} O(n^{\alpha(p-1)-1})\ell(n)du
$$

$$
\leq C\int_0^1 Q_{\|X\|}^p(u)\ell(Q_{\|X\|}^{1/\alpha}(u))du
$$

$$
= C\mathbb{E}[\|X\|^p\ell(\|X\|^{1/\alpha})]. \tag{1.28}
$$

Next we will control the first term on right hand in (1.22). Since

$$\left\{\left(\left\|\sum_{j=1}^{k}A_{nj}Y'_{nj}\right\|,\mathcal{F}_{nk}\right),1\leq k\leq n,n\geq 1\right\}$$

is a triangular array of real-valued submartingales, applying Doob's inequality to

$$\left\{\left(\left\|\sum_{j=1}^{k}A_{nj}Y'_{nj}\right\|,\mathcal{F}_{nk}\right),1\leq k\leq n,n\geq 1\right\},$$

we obtain that

$$\mathbb{P}\left\{\sup_{1\leq k\leq n}\left\|\sum_{j=1}^{k}A_{nj}Y'_{nj}\right\|\geq\varepsilon n^{\alpha}\right\}\leq\frac{1}{\varepsilon^{r}n^{r\alpha}}\mathbb{E}\left\|\sum_{j=1}^{n}A_{nj}Y'_{nj}\right\|^{r}.$$

Since $\mathbb{B}$ is r-smooth for some $r>p$, by applying Inequality (1.14), we have that

$$\mathbb{P}\left\{\sup_{1\leq k\leq n}\left\|\sum_{j=1}^{k}A_{nj}Y'_{nj}\right\|\geq\varepsilon n^{\alpha}\right\}\leq\frac{D}{\varepsilon^{r}n^{r\alpha}}\sum_{j=1}^{n}\mathbb{E}\|A_{nj}Y'_{nj}\|^{r}$$

$$\leq\frac{2^{r}D}{\varepsilon^{r}n^{r\alpha}}\sum_{j=1}^{n}\mathbb{E}\|A_{nj}X'_{nj}\|^{r}.$$

Applying the independence of A_{nj} and X_{nj} for any $n\geq 1$ and $1\leq j\leq n$, and the definition of X'_{nj}, we obtain that

$$\mathbb{P}\left\{\sup_{1\leq k\leq n}\left\|\sum_{j=1}^{k}A_{nj}Y'_{nj}\right\|\geq\varepsilon n^{\alpha}\right\}$$

$$\leq\frac{2^{r}D}{\varepsilon^{r}n^{r\alpha}}\sum_{j=1}^{n}\mathbb{E}|A_{nj}|^{r}\mathbb{E}\|X'_{nj}\|^{r}$$

$$=\frac{2^{r}D}{\varepsilon^{r}n^{r\alpha}}\sum_{j=1}^{n}\mathbb{E}|A_{nj}|^{r}\mathbb{E}\left(\|X_{nj}\|^{r}\mathbf{1}_{\{\|X_{nj}\|\leq n^{\alpha}\}}\right)$$

$$=\frac{2^{r}D}{\varepsilon^{r}n^{r\alpha}}\sum_{j=1}^{n}\left(\mathbb{E}|A_{nj}|^{r}\int_{0}^{1}Q^{r}_{\|X_{nj}\|\mathbf{1}_{\{\|X_{nj}\|\leq n^{\alpha}\}}}(u)\mathrm{d}u\right).\qquad(1.29)$$

Noting that for any $A>0$, any $n\geq 1$ and $1\leq j\leq n$, $\|X_{nj}\|\mathbf{1}_{\{\|X_{nj}\|\leq A\}}\leq\|X_{nj}\|\wedge A$. Hence, for any $u\in[0,1]$ and $n\geq 1$,

$$Q_{\|X_{nj}\|\mathbf{1}_{\{\|X_{nj}\|\leq A\}}}(u)\leq Q_{\|X_{nj}\|\wedge A}(u)\leq Q_{\|X\|\wedge A}(u)\leq Q_{\|X\|}(u)\wedge A.\qquad(1.30)$$

From (1.29), (1.30), (1.16) and the fact that $A < Q_{\|X\|}(u)$ if and only if $u < L_{\|X\|}(A)$, we have that

$$\sum_{n=1}^{\infty} n^{\alpha p-2}\ell(n)\mathbb{P}\left\{\sup_{1\le k\le n}\left\|\sum_{j=1}^{k}A_{nj}Y'_{nj}\right\| \ge \varepsilon n^{\alpha}\right\}$$

$$\le \frac{2^r D}{\varepsilon^r}\sum_{n=1}^{\infty}O(n^{\alpha(p-r)-1})\ell(n)\int_0^1 \left(Q_{\|X\|}(u)\wedge n^{\alpha}\right)^r du$$

$$\le \frac{2^r D}{\varepsilon^r}(A_1(n)+A_2(n)), \tag{1.31}$$

where

$$A_1(n)=\sum_{n=1}^{\infty}O(n^{\alpha(p-r)-1})\ell(n)\int_0^1 n^{\alpha r}\mathbf{1}_{\{u<L_{\|X\|}(n^{\alpha})\}}du,$$

$$A_2(n)=\sum_{n=1}^{\infty}O(n^{\alpha(p-r)-1})\ell(n)\int_0^1 Q_{\|X\|}^r(u)\mathbf{1}_{\{L_{\|X\|}(n^{\alpha})\le u\le 1\}}du.$$

Since $u \ge L_{\|X\|}(n^{\alpha})$ if and only if $n \ge Q_{\|X\|}^{1/\alpha}(u)$, by Lemmas 1.1 and 1.2, there exists two finite constants C_1 and C_2 depending only on α, p and r, such that

$$A_1(n)\le \sum_{n=1}^{\infty}O(n^{\alpha(p+1-r)-1})\ell(n)\int_0^1 Q_{\|X\|}^{r-1}(u)\mathbf{1}_{\{u<L_{\|X\|}(n^{\alpha})\}}du$$

$$\le \int_0^1 Q_{\|X\|}^{r-1}(u)\sum_{n=1}^{[Q_{\|X\|}^{1/\alpha}(u)]}O(n^{\alpha(p+1-r)-1})\ell(n)du$$

$$\le C_1 \int_0^1 Q_{\|X\|}^p(u)\ell(Q_{\|X\|}^{1/\alpha}(u))du$$

$$= C_1\mathbb{E}[\|X\|^p\ell(\|X\|^{1/\alpha})], \tag{1.32}$$

$$A_2(n)\le \int_0^1 Q_{\|X\|}^r(u)\sum_{n\ge Q_{\|X\|}^{1/\alpha}(u)}O(n^{\alpha(p-r)-1})\ell(n)du$$

$$\le C_2 \int_0^1 Q_{\|X\|}^p(u)\ell(Q_{\|X\|}^{1/\alpha}(u))du$$

$$= C_2\mathbb{E}[\|X\|^p\ell(\|X\|^{1/\alpha})]. \tag{1.33}$$

Thus $A_1(n)$ and $A_2(n)$ are finite as soon as $\mathbb{E}\left[\|X\|^p\ell(\|X\|^{1/\alpha})\right]$ is finite. Therefore (a) holds.

(b) Since $\mathbb{B}$ is super-reflexive, it is r-smooth for some $1 < r \le 2$. Without loss of generality, we can assume that $r < 2$. Using Inequality (1.24) with $\alpha = 1$, for all $n \geq 1$ and $1 \leq j \leq n$,

$$\mathbb{E}\left(\|X_{nj}\|1_{\{\|X_{nj}\|>n\}}\right) \leq \mathbb{E}\left(\|X\|1_{\{\|X\|>n\}}\right). \tag{1.34}$$

Therefore, by applying Fubini Theorem, we see from (1.23) with $\alpha = 1$ that if $\mathbb{E}\left[\|X\|\ln^+\|X\|\right] < +\infty$ and (1.18) holds, then

$$\sum_{n=1}^{\infty} n^{-1}\mathbb{P}\left\{\sup_{1\leq k\leq n}\left\|\sum_{j=1}^{k} A_{nj}Y''_{nj}\right\| \geq \varepsilon n\right\} \leq \frac{2O(1)}{\varepsilon}\sum_{n=1}^{\infty} n^{-1}\mathbb{E}\left(\|X\|1_{\{\|X\|>n\}}\right)$$

$$= \frac{2O(1)}{\varepsilon}\mathbb{E}\left(\|X\|\sum_{1\leq n\leq[\|X\|]} n^{-1}\right)$$

$$\leq \frac{2O(1)}{\varepsilon}\mathbb{E}\left[\|X\|\ln^+\|X\|\right]$$

$$< \infty. \tag{1.35}$$

Finally, we will prove that $\sum_{n=1}^{\infty} n^{-1}\mathbb{P}\{\sup_{1\leq k\leq n}\|\sum_{j=1}^{k} A_{nj}Y'_{nj}\| \geq \varepsilon n\} < \infty$. Starting from (1.31) with $\alpha = p = 1$ and $\ell(\cdot) = 1$, and applying Inequalities (1.32) and (1.33) with $\alpha = p = 1$ and $\ell(\cdot) = 1$, we obtain that $A_1(n)$ and $A_2(n)$ are finite as soon as $\mathbb{E}\|X\|$ is finite, which completes the proof of (b). ∎

Chapter 2

C.M.C. for R.W.S. of T.A. of $\mathbb{B}$-valued M.D.

The full name of the title of Chapter 2 is complete moment convergence for randomly weighted sums of the triangular array of Banach space valued martingale differences.

It is well known that partial sum is a particular case of weighted sum. It should be noted that many linear models in statistics based on a random sample involve weighted sums of dependent random variables. Examples include least-squares estimators, nonparametric regression function estimators, and jackknife estimates, among others. In this respect, the study of convergence for these weighted sums has impact on statistics, probability, and their applications. (cf. [133]) In fact, the study of convergence for these weighted sums also has impact on systems theory, mathematical physics, applied economics, and their applications. Thus the convergence for the weighted sums seems more important.

Starting from the 1970s, the random nature of many problems arising in the applied sciences is noted. This leads to mathematical models which deal with the limiting behaviour of weighted sums of random elements in normed linear spaces, where the weights are random variables. Taylor (1972, [127]) and Taylor and Padgett (1974, [129]; 1976, [130]) obtained some basic results by considering random weights $(A_j)_{j \geq 1}$. From 1978 on, it begins to be studied directly the convergence of randomly weighted sums of random elements in separable Banach spaces or in separable general normed linear spaces, such as Wei and Taylor (1978, [145, 146]), Taylor and Calhoun (1983, [128]), Taylor et al. (1984, [131]), Ordóñez Cabrera (1988, [109]), Adler et al. (1992, [2]), Wang and Rao (1995, [139]) and Hu and Chang (1999, [51]). The limiting behaviour of randomly weighted sums plays an important role in various applied and theoretical problems. One can see the example of Rosalsky and Sreehari (1998, [114]), in queueing theory. (cf. [55])

Chow (1988, [26]) first studied the complete moment convergence, which is more exact than the complete convergence and implies the complete convergence. Recently, Wang and Hu (2014, [141]) investigated the complete moment convergence for the maximal partial sums of the sequence of real-valued martingale differences; Yang *et al.* (2013, [147]) obtained the complete moment convergence for single-indexed randomly weighted sums of the sequence of real-valued martingale differences, Yang *et al.* (2014, [149]) considered the complete moment convergence of double-indexed weighted sums of the sequence of real-valued martingale differences.

For one thing, most literature in Banach space setting is about the weighted sums of the (triangular) array of rowwise independent Banach space valued random elements, few literature (cf. [42]) is about the weighted sums of the (triangular) array of Banach space valued martingale differences. For another thing, some results of Hao (2013, [42]) required that the martingale differences are identically distributed, which is too strong to apply conveniently. Furthermore, a number of researchers considered the weighted sums of Banach space valued random elements where the weights are real constants. To the best of our knowledge, there has only been few results (cf. [133]) in the literature on the complete (moment) convergence for weighted sums of the array of Banach space valued random elements in which the weights are real-valued random variables. Thus we consider the complete moment convergence for double-indexed randomly weighted sums of the triangular array of Banach space valued martingale differences which implies their complete convergence. The new challenge is that one has to treat a weighted sum of the triangular array of Banach space valued martingale differences whose weights are random with double indices.

Our main objective is twofold. One is to establish the complete moment convergence for double-indexed randomly weighted sums of the triangular array of Banach space valued martingale differences. The other is to extend Theorem 5 of Yang *et al.* (2014, [149]), Theorems 3.1, 3.3 and 3.4 of Wang and Hu (2014, [141]), Theorems 2.1, 2.2 and 2.3 of Yang *et al.* (2013, [147]), and Theorems 1.4 and 1.6 of Wang *et al.* (2012, [142]) to double-indexed randomly weighted sums of the triangular array of Banach space valued martingale differences.

The rest of the chapter is organized as follows. In Section 2.1 we mainly consider the complete moment convergence for double-indexed randomly weighted sums of the triangular array of Banach space valued martingale differences, and extend Theorem 5 of Yang *et al.* (2014, [149]), Theorems 3.1, 3.3 and 3.4 of Wang and Hu (2014, [141]), Theorems 2.1, 2.2 and 2.3 of Yang *et al.* (2013, [147]) and Theorems 1.4 and 1.6 of Wang *et al.* (2012, [142]). Section 2.2 is devoted to some preliminary facts which are used in Section 2.3, where the proofs of the main results are given.

2.1 Main results

In this section, we mainly consider the complete moment convergence for double-indexed randomly weighted sums of the triangular array of Banach space valued martingale differences.

Let $(\Omega, \mathcal{F}, \mathbb{P})$ be a probability space and $(\mathbb{B}, \|\cdot\|)$ be a separable Banach space.

Let $\{(X_{nj}, \mathcal{F}_{nj}), 1 \leq j \leq n, n \geq 1\}$ be a triangular array of $\mathbb{B}$-valued martingale differences.

Definition 2.1 *We say that a triangular array of $\mathbb{B}$-valued random elements $\{X_{nj}, 1 \leq j \leq n, n \geq 1\}$ is stochastically dominated by a $\mathbb{B}$-valued random element X if for some constant $C < \infty$,*

$$\mathbb{P}\{\|X_{nj}\| > t\} \leq C\mathbb{P}\{\|X\| > t\}, \ t \geq 0, \ n \geq 1, \ 1 \leq j \leq n.$$

In the following theorems $\ell(\cdot) > 0$ is a function slowly varying at ∞.

Theorem 5 of Yang *et al.* (2014, [149]), Theorems 3.1, 3.3 and 3.4 of Wang and Hu (2014, [141]), Theorems 2.1, 2.2 and 2.3 of Yang *et al.* (2013, [147]), and Theorems 1.4 and 1.6 of Wang *et al.* (2012, [142]) are some nice results, nevertheless in applications instead of a real-valued single martingale we often need to consider a $\mathbb{B}$-valued martingale array: for example when we use the decomposition of a random sequence (S_n) into $\mathbb{B}$-valued martingale differences (such as in the study of directed polymers in a random environment), the summands usually depend on n: $S_n = \sum_{j=1}^{n} X_{nj}$, $X_{nj} = \mathbb{E}[S_n|\mathcal{F}_j] - \mathbb{E}[S_n|\mathcal{F}_{j-1}]$, where $\mathcal{F}_0 = \{\emptyset, \Omega\}$ and $\mathcal{F}_i = \sigma(S_1, \cdots, S_i)$ for $i \geq 1$. In the following we give some sufficient conditions for the complete moment convergence and the complete convergence of double-indexed randomly weighted sums of the triangular array $\{(X_{nj}, \mathcal{F}_{nj}), 1 \leq j \leq n, n \geq 1\}$ of Banach space valued martingale differences and extend Theorem 5 of Yang *et al.* (2014, [149]), Theorems 3.1, 3.3 and 3.4 of Wang and Hu (2014, [141]), Theorems 2.1, 2.2 and 2.3 of Yang *et al.* (2013, [147]), and Theorems 1.4 and 1.6 of Wang *et al.* (2012, [142]) to double-indexed randomly weighted sums of the triangular array of Banach space valued martingale differences.

Theorem 2.1 *Let $1 < p < 2$, $1/p \leq \alpha < 2/p$ and $\{(X_{nj}, \mathcal{G}_{nj}), 1 \leq j \leq n, n \geq 1\}$ be a triangular array of $\mathbb{B}$-valued martingale differences stochastically dominated by a $\mathbb{B}$-valued random element X with*

$$\mathbb{E}[\|X\|^p \ell(\|X\|^{1/\alpha})] < \infty,$$

where $\mathcal{G}_{nj} = \sigma(X_{n1}, X_{n2}, \cdots, X_{nj})$, $1 \leq j \leq n$, $n \geq 1$, and $\mathcal{G}_{n0} = \{\emptyset, \Omega\}$. Let $\{A_{nj}, 1 \leq j \leq n, n \geq 1\}$ be a triangular array of random variables, which is *independent of* $\{X_{nj}, 1 \le j \le n, n \ge 1\}$. *If*

$$\sum_{j=1}^{n} \mathbb{E}A_{nj}^2 = O(n), \qquad (2.1)$$

then

$$\sum_{n=1}^{\infty} n^{p\alpha-2-\alpha}\ell(n)\mathbb{E}\left(\max_{1\le k\le n}\left\|\sum_{j=1}^{k} A_{nj}X_{nj}\right\| - \varepsilon n^{\alpha}\right)^{+} < \infty \qquad (2.2)$$

and

$$\sum_{n=1}^{\infty} n^{p\alpha-2}\ell(n)\mathbb{P}\left\{\max_{1\le k\le n}\left\|\sum_{j=1}^{k} A_{nj}X_{nj}\right\| > \varepsilon n^{\alpha}\right\} < \infty \qquad (2.3)$$

for all $\varepsilon > 0$.

For a sequence of single real-valued martingale differences, in the case where $A_{nj} = A_j$ for all $n \ge 1$ and $1 \le j \le n$, X is a nonnegative random variable and $\ell(\cdot) = 1$, by Theorem 2.1, we get Theorem 2.1 of Yang *et al.* (2013, [147]).

Theorem 2.2 *Let* $p \ge 2$, $\alpha > 1/2$ *and* $\{(X_{nj}, \mathcal{G}_{nj}), 1 \le j \le n, n \ge 1\}$ *be a triangular array of* B-*valued martingale differences stochastically dominated by a* B-*valued random element* X *with*

$$\mathbb{E}[\|X\|^p\ell(\|X\|^{1/\alpha})] < \infty,$$

where $\mathcal{G}_{nj} = \sigma(X_{n1}, X_{n2}, \cdots, X_{nj})$, $1 \le j \le n$, $n \ge 1$, *and* $\mathcal{G}_{n0} = \{\emptyset, \Omega\}$. *Let* $\{A_{nj}, 1 \le j \le n, n \ge 1\}$ *be a triangular array of random variables, which is independent of* $\{X_{nj}, 1 \le j \le n, n \ge 1\}$. *For some* $q > 2(\alpha p - 1)/(2\alpha - 1)$, *assume that* $\mathbb{E}\{\sup_{n,j:1\le j\le n, n\ge 1} \mathbb{E}[\|X_{nj}\|^2 \,|\mathcal{G}_{n,j-1}]\}^{q/2} < \infty$ *and*

$$\sum_{j=1}^{n} \mathbb{E}|A_{nj}|^q = O(n), \qquad (2.4)$$

then (2.2) *and* (2.3) *hold for all* $\varepsilon > 0$.

For a sequence of single real-valued martingale differences, in the case where $\{A_{nj}, 1 \le j \le n, n \ge 1\}$ is a triangular array of real numbers, X is a nonnegative random variable and $\ell(\cdot) = 1$, by Theorem 2.2, we get Theorem 5 of Yang *et al.* (2014, [149]); in the case where $A_{nj} = A_j$ for all $n \ge 1$ and $1 \le j \le n$, X is a nonnegative random variable and $\ell(\cdot) = 1$, by Theorem 2.2, we get Theorem 2.2 of Yang *et al.* (2013, [147]); in the case where $A_{nj} = 1$ for all $n \ge 1$ and $1 \le j \le n$, X is a random variable, by Theorems 2.1 and 2.2, we get Theorems 3.1 and 3.3 of Wang and Hu (2014, [141]) for $1 < p < 2$ and $1 \le \alpha p < 2$, or $\alpha > 1/2$ and $p \ge 2$;

in the case where $A_{nj} = 1$ for all $n \geq 1$ and $1 \leq j \leq n$, X is a random variable, and $\sup_{1 \leq j \leq n, n \geq 1} \mathbb{E}[\|X_{nj}\|^2 |\mathcal{G}_{n,j-1}] \leq C$ a.s. if $p \geq 2$, by Theorems 2.1 and 2.2, we get Theorem 1.4 of Wang et al. (2012, [142]) for $1 < p < 2$ and $1 \leq \alpha p < 2$, or $\alpha > 1/2$ and $p \geq 2$.

Let $\ell(\cdot) = 1$ and $p = 2/\alpha$ for $1/2 < \alpha \leq 1$ in Theorem 2.2, we get the following corollary.

Corollary 2.1 *Let $1/2 < \alpha \leq 1$ and $\{(X_{nj}, \mathcal{G}_{nj}), 1 \leq j \leq n, n \geq 1\}$ be a triangular array of $\mathbb{B}$-valued martingale differences stochastically dominated by a $\mathbb{B}$-valued random element X with*

$$\mathbb{E}\|X\|^{2/\alpha} < \infty,$$

where $\mathcal{G}_{nj} = \sigma(X_{n1}, X_{n2}, \cdots, X_{nj})$, $1 \leq j \leq n$, $n \geq 1$, and $\mathcal{G}_{n0} = \{\emptyset, \Omega\}$. Let $\{A_{nj}, 1 \leq j \leq n, n \geq 1\}$ be a triangular array of random variables, which is independent of $\{X_{nj}, 1 \leq j \leq n, n \geq 1\}$. For some $q > 2/(2\alpha - 1)$, assume that $\mathbb{E}\{\sup_{n,j:1 \leq j \leq n, n \geq 1} \mathbb{E}[\|X_{nj}\|^2 |\mathcal{G}_{n,j-1}]\}^{q/2} < \infty$ and (2.4) holds, then

$$\sum_{n=1}^{\infty} n^{-\alpha} \mathbb{E}\left(\max_{1 \leq k \leq n} \left\| \sum_{j=1}^{k} A_{nj} X_{nj} \right\| - \varepsilon n^\alpha \right)^+ < \infty \tag{2.5}$$

and

$$\sum_{n=1}^{\infty} \mathbb{P}\left\{ \max_{1 \leq k \leq n} \left\| \sum_{j=1}^{k} A_{nj} X_{nj} \right\| > \varepsilon n^\alpha \right\} < \infty \tag{2.6}$$

for all $\varepsilon > 0$. In particular, one has

$$\lim_{n \to \infty} \frac{1}{n^\alpha} \sum_{j=1}^{n} A_{nj} X_{nj} = 0, \quad a.s. \tag{2.7}$$

Remark 2.3 *Corollary 2.1 can be applied to study the convergence of the state observers of linear-time-invariant systems. (cf. [138, 134, 149])*

Theorem 2.4 *Let $\alpha > 0$ and $\{(X_{nj}, \mathcal{G}_{nj}), 1 \leq j \leq n, n \geq 1\}$ be a triangular array of $\mathbb{B}$-valued martingale differences stochastically dominated by a $\mathbb{B}$-valued random element X with*

$$\mathbb{E}(\|X\| \ln^+ \|X\|) < \infty,$$

where $\mathcal{G}_{nj} = \sigma(X_{n1}, X_{n2}, \cdots, X_{nj})$, $1 \leq j \leq n$, $n \geq 1$, and $\mathcal{G}_{n0} = \{\emptyset, \Omega\}$. Let $\{A_{nj}, 1 \leq j \leq n, n \geq 1\}$ be a triangular array of random variables, which is independent of $\{X_{nj}, 1 \leq j \leq n, n \geq 1\}$. If (2.1) holds, then

$$\sum_{n=1}^{\infty} n^{-2} \mathbb{E}\left(\max_{1 \leq k \leq n} \left\| \sum_{j=1}^{k} A_{nj} X_{nj} \right\| - \varepsilon n^\alpha \right)^+ < \infty \tag{2.8}$$

and

$$\sum_{n=1}^{\infty} n^{\alpha-2} \mathbb{P} \left\{ \max_{1 \leq k \leq n} \left\| \sum_{j=1}^{k} A_{nj} X_{nj} \right\| > \varepsilon n^{\alpha} \right\} < \infty \qquad (2.9)$$

for all $\varepsilon > 0$.

For a sequence of single real-valued martingale differences, in the case where $A_{nj} = A_j$ for all $n \geq 1$ and $1 \leq j \leq n$, and X is a nonnegative random variable, by Theorem 2.4, we get Theorem 2.3 of Yang *et al.* (2013, [147]); in the case where $\alpha \geq 1$ and $A_{nj} = 1$ for all $n \geq 1$ and $1 \leq j \leq n$, and X is a random variable, by Theorem 2.4, we get Theorem 3.4 of Wang and Hu (2014, [141]) and Theorem 1.6 of Wang *et al.* (2012, [142]).   ∎

2.2   Preliminaries

The proofs of the theorems in Section 2.2 will need the following five lemmas.

Lemma 2.1 (*cf. Hall and Heyde [40], Theorem 2.11*) *If $\{(X_j, \mathcal{F}_j)\}_{j=1}^{n}$ be a sequence of $\mathbb{B}$-valued martingale differences and $p > 0$, then there exists a constant C depending only on p such that for $n \geq 1$,*

$$\mathbb{E}\left(\max_{1 \leq k \leq n} \left\| \sum_{j=1}^{k} X_j \right\|^p \right) \leq C \left\{ \mathbb{E}\left(\sum_{j=1}^{n} \mathbb{E}[\|X_j\|^2 | \mathcal{F}_{j-1}] \right)^{p/2} + \mathbb{E}\left(\max_{1 \leq j \leq n} \|X_j\|^p \right) \right\}.$$

Lemma 2.2 (*cf. Sung [120], Lemma 2.4*) *Let $(X_n)_{n \geq 1}$ and $(Y_n)_{n \geq 1}$ be sequences of $\mathbb{B}$-valued random elements. Then for any $n \geq 1$, $q > 1$, $\varepsilon > 0$ and $a > 0$,*

$$\mathbb{E}\left(\max_{1 \leq k \leq n} \left\| \sum_{j=1}^{k} (X_j + Y_j) \right\| - \varepsilon a \right)^{+} \leq \left(\frac{1}{\varepsilon^q} + \frac{1}{q-1} \right) \frac{1}{a^{q-1}} \mathbb{E}\left(\max_{1 \leq k \leq n} \left\| \sum_{j=1}^{k} X_j \right\|^q \right)$$

$$+ \mathbb{E}\left(\max_{1 \leq k \leq n} \left\| \sum_{j=1}^{k} Y_j \right\| \right).$$

Lemma 2.3 (*cf. Adler et al. [1], Lemma 1 and Lemma 3, or Wang et al. [142], Lemma 2.2*) *Let $(X_n)_{n \geq 1}$ be a sequence of $\mathbb{B}$-valued random elements stochastically dominated by a $\mathbb{B}$-valued random element X. Then for any $n \geq 1$, $a > 0$ and $b > 0$,*

$$\mathbb{E}[\|X_n\|^a \mathbf{1}_{\{\|X_n\| \leq b\}}] \leq C_1 \left\{ \mathbb{E}[\|X\|^a \mathbf{1}_{\{\|X\| \leq b\}}] + b^a \mathbb{P}\{\|X\| > b\} \right\}$$

and

$$\mathbb{E}[\|X_n\|^a \mathbf{1}_{\{\|X_n\| > b\}}] \leq C_2 \mathbb{E}[\|X\|^a \mathbf{1}_{\{\|X\| > b\}}].$$

Consequently, $\mathbb{E}\|X_n\|^a \leq C_3 \mathbb{E}\|X\|^a$. Here C_1, C_2 and C_3 are positive constants.

The proofs of Lemmas 2.1-2.3 are similar to those of the corresponding lemmas in references.

Lemma 2.4 (*cf. Hao [42], Lemma 23*) *Let* $\alpha > -1$. *Then for some* n_0, c_1, $c_2 > 0$ *and all* $N \geq n_0$,

$$c_1 N^{\alpha+1}\ell(N) \leq \sum_{n=1}^{N} n^\alpha \ell(n) \leq c_2 N^{\alpha+1}\ell(N).$$

Lemma 2.5 *Let* $\alpha < -1$. *Then for some* n_3, c_4, $c_5 > 0$ *and all* $N \geq n_3$,

$$c_4 N^{\alpha+1}\ell(N) \leq \sum_{n=N}^{\infty} n^\alpha \ell(n) \leq c_5 N^{\alpha+1}\ell(N).$$

2.3 Proofs of the main results

We get Theorem 2.1 by Lemmas 2.1, 2.2, 2.3, 2.4 and 2.5.
Proof of Theorem 2.1. Let

$$\mathcal{G}_{n0} = \{\emptyset, \Omega\},$$

$$\mathcal{G}_{nj} = \sigma(X_{n1}, \ X_{n2}, \ \cdots, \ X_{nj}),$$

$$X'_{nj} = X_{nj}\mathbf{1}_{\{\|X_{nj}\| \leq n^\alpha\}}$$

and

$$X''_{nj} = X_{nj}\mathbf{1}_{\{\|X_{nj}\| > n^\alpha\}}$$

for $1 \leq j \leq n$ and $n \geq 1$. Then

$$
\begin{aligned}
A_{nj}X_{nj} ={}& A_{nj}X_{nj}\mathbf{1}_{\{\|X_{nj}\| > n^\alpha\}} \\
& + \left\{ A_{nj}X'_{nj} - \mathbb{E}\left[A_{nj}X'_{nj} \big| \mathcal{G}_{n,j-1} \right] \right\} \\
& + \mathbb{E}\left[A_{nj}X'_{nj} \big| \mathcal{G}_{n,j-1} \right]
\end{aligned}
\tag{2.10}
$$

for $1 \le j \le n$ and $n \ge 1$. By Lemma 2.2 with $a = n^\alpha$, for $q > 1$, we have that

$$
\sum_{n=1}^{\infty} n^{p\alpha-2-\alpha} \ell(n) \mathbb{E} \left(\max_{1 \le k \le n} \left\| \sum_{j=1}^{k} A_{nj} X_{nj} \right\| - \varepsilon n^\alpha \right)^{+}
$$

$$
\le C_1 \sum_{n=1}^{\infty} n^{p\alpha-2-q\alpha} \ell(n) \mathbb{E} \left(\max_{1 \le k \le n} \left\| \sum_{j=1}^{k} \left\{ A_{nj} X'_{nj} - \mathbb{E}[A_{nj} X'_{nj}|\mathcal{G}_{n,j-1}] \right\} \right\|^{q} \right)
$$

$$
+ \sum_{n=1}^{\infty} n^{p\alpha-2-\alpha} \ell(n) \mathbb{E} \left(\max_{1 \le k \le n} \left\| \sum_{j=1}^{k} \left\{ A_{nj} X''_{nj} + \mathbb{E}[A_{nj} X'_{nj}|\mathcal{G}_{n,j-1}] \right\} \right\| \right)
$$

$$
\le C_1 \sum_{n=1}^{\infty} n^{p\alpha-2-q\alpha} \ell(n) \mathbb{E} \left(\max_{1 \le k \le n} \left\| \sum_{j=1}^{k} \left\{ A_{nj} X'_{nj} - \mathbb{E}[A_{nj} X'_{nj}|\mathcal{G}_{n,j-1}] \right\} \right\|^{q} \right)
$$

$$
+ \sum_{n=1}^{\infty} n^{p\alpha-2-\alpha} \ell(n) \mathbb{E} \left(\max_{1 \le k \le n} \left\| \sum_{j=1}^{k} A_{nj} X''_{nj} \right\| \right)
$$

$$
+ \sum_{n=1}^{\infty} n^{p\alpha-2-\alpha} \ell(n) \mathbb{E} \left(\max_{1 \le k \le n} \left\| \sum_{j=1}^{k} \mathbb{E}[A_{nj} X'_{nj}|\mathcal{G}_{n,j-1}] \right\| \right)
$$

$$
:= C_1 H_1 + H_2 + H_3. \tag{2.11}
$$

By Hölder's inequality and (2.1), we see that

$$
\sum_{j=1}^{n} \mathbb{E}|A_{nj}|
$$

$$
\le \left(\sum_{j=1}^{n} \mathbb{E} A_{nj}^2 \right)^{1/2} \left(\sum_{j=1}^{n} 1 \right)^{1/2}
$$

$$
= O(n). \tag{2.12}
$$

Since $\{A_{nj}, 1 \le j \le n, n \ge 1\}$ is independent of $\{X_{nj}, 1 \le j \le n, n \ge 1\}$, by Lemma 2.3, (2.12), Lemma 2.4 and $\mathbb{E}[\|X\|^p \ell(\|X\|^{1/\alpha})] < \infty$, we see that

$$
H_2 = \sum_{n=1}^{\infty} n^{\alpha p-2-\alpha} \ell(n) \mathbb{E} \left(\max_{1 \le k \le n} \left\| \sum_{j=1}^{k} A_{nj} X_{nj} \mathbf{1}_{\{\|X_{nj}\| > n^\alpha\}} \right\| \right)
$$

$$
\le \sum_{n=1}^{\infty} n^{\alpha p-2-\alpha} \ell(n) \sum_{j=1}^{n} \mathbb{E}|A_{nj}| \mathbb{E} \left[\|X_{nj}\| \mathbf{1}_{\{\|X_{nj}\| > n^\alpha\}} \right]
$$

$$\leq C_2 \sum_{n=1}^{\infty} n^{\alpha p-1-\alpha} \ell(n) \mathbb{E}\left[\|X\| \mathbf{1}_{\{\|X\|>n^{\alpha}\}}\right]$$

$$= C_2 \sum_{n=1}^{\infty} n^{\alpha p-1-\alpha} \ell(n) \sum_{m=n}^{\infty} \mathbb{E}\left[\|X\| \mathbf{1}_{\{m^{\alpha}<\|X\|\leq(m+1)^{\alpha}\}}\right]$$

$$= C_2 \sum_{m=1}^{\infty} \mathbb{E}\left[\|X\| \mathbf{1}_{\{m^{\alpha}<\|X\|\leq(m+1)^{\alpha}\}}\right] \sum_{n=1}^{m} n^{\alpha p-1-\alpha} \ell(n)$$

$$\leq C_3 \sum_{m=1}^{\infty} m^{\alpha p-\alpha} \ell(m) \mathbb{E}\left[X \mathbf{1}_{\{m^{\alpha}<\|X\|\leq(m+1)^{\alpha}\}}\right]$$

$$\leq C_4 \mathbb{E}\left[\|X\|^p \ell\left(\|X\|^{1/\alpha}\right) \sum_{m=1}^{\infty} \mathbf{1}_{\{m^{\alpha}<\|X\|\leq(m+1)^{\alpha}\}}\right]$$

$$= C_4 \mathbb{E}\left[\|X\|^p \ell\left(\|X\|^{1/\alpha}\right) \mathbf{1}_{\{\|X\|>1\}}\right]$$

$$\leq C_4 \mathbb{E}\left[\|X\|^p \ell\left(\|X\|^{1/\alpha}\right)\right]$$

$$< \infty. \tag{2.13}$$

Next, we will prove $H_3 < \infty$. Since $\{(X_{nj}, \mathcal{G}_{nj}), 1 \leq j \leq n, n \geq 1\}$ is a triangular array of $\mathbb{B}$-valued martingale differences and $\{A_{nj}, 1 \leq j \leq n, n \geq 1\}$ is independent of $\{X_{nj}, 1 \leq j \leq n, n \geq 1\}$, we have that

$$\mathbb{E}\left[A_{nj}X_{nj}\big|\mathcal{G}_{n,j-1}\right]$$
$$= \mathbb{E}A_{nj}\mathbb{E}\left[X_{nj}\big|\mathcal{G}_{n,j-1}\right]$$
$$= 0 \quad \text{a.s.}$$

for $1 \leq j \leq n$ and $n \geq 1$. Consequently, by the proof of (2.13), we obtain that

$$H_3 = \sum_{n=1}^{\infty} n^{\alpha p-2-\alpha} \ell(n) \mathbb{E}\left(\max_{1\leq k\leq n}\left\|\sum_{j=1}^{k} \mathbb{E}[A_{nj}X'_{nj}|\mathcal{G}_{n,j-1}]\right\|\right)$$

$$= \sum_{n=1}^{\infty} n^{\alpha p-2-\alpha} \ell(n) \mathbb{E}\left(\max_{1\leq k\leq n}\left\|\sum_{j=1}^{k} \mathbb{E}\left[A_{nj}X_{nj}\mathbf{1}_{\{\|X_{nj}\|\leq n^{\alpha}\}}\big|\mathcal{G}_{n,j-1}\right]\right\|\right)$$

$$= \sum_{n=1}^{\infty} n^{\alpha p-2-\alpha} \ell(n) \mathbb{E}\left(\max_{1\leq k\leq n}\left\|\sum_{j=1}^{k} \mathbb{E}\left[A_{nj}X_{nj}\mathbf{1}_{\{\|X_{nj}\|>n^{\alpha}\}}\big|\mathcal{G}_{n,j-1}\right]\right\|\right)$$

$$\leq \sum_{n=1}^{\infty} n^{\alpha p-2-\alpha} \ell(n) \sum_{j=1}^{n} \mathbb{E}|A_{nj}|\mathbb{E}\left[\|X_{nj}\|\mathbf{1}_{\{\|X_{nj}\|>n^{\alpha}\}}\right]$$

$$\leq C_2 \sum_{n=1}^{\infty} n^{\alpha p-1-\alpha} \ell(n) \mathbb{E}\left[\|X\| \mathbf{1}_{\{\|X\|>n^{\alpha}\}}\right]$$

$$= C_2 \sum_{n=1}^{\infty} n^{\alpha p-1-\alpha} \ell(n) \sum_{m=n}^{\infty} \mathbb{E}\left[\|X\| \mathbf{1}_{\{m^{\alpha}<\|X\|\leq(m+1)^{\alpha}\}}\right]$$

$$= C_2 \sum_{m=1}^{\infty} \mathbb{E}\left[\|X\| \mathbf{1}_{\{m^{\alpha}<\|X\|\leq(m+1)^{\alpha}\}}\right] \sum_{n=1}^{m} n^{\alpha p-1-\alpha} \ell(n)$$

$$\leq C_3 \sum_{m=1}^{\infty} m^{\alpha p-\alpha} \ell(m) \mathbb{E}\left[X \mathbf{1}_{\{m^{\alpha}<\|X\|\leq(m+1)^{\alpha}\}}\right]$$

$$\leq C_4 \mathbb{E}\left[\|X\|^p \ell\left(\|X\|^{1/\alpha}\right) \sum_{m=1}^{\infty} \mathbf{1}_{\{m^{\alpha}<\|X\|\leq(m+1)^{\alpha}\}}\right]$$

$$\leq C_4 \mathbb{E}\left[\|X\|^p \ell\left(\|X\|^{1/\alpha}\right)\right]$$

$$< \infty. \tag{2.14}$$

Finally, we turn to prove $H_1 < \infty$. Since $\{(a_{nj}X'_{nj}-\mathbb{E}[a_{nj}X'_{nj}|\mathcal{G}_{n,j-1}], \mathcal{G}_{nj})\}_{j=1}^n$ is a martingale difference sequence for fixed real numbers $a_{n1}, a_{n2}, \cdots, a_{nn}$ and $\{A_{nj}, 1 \leq j \leq n, n \geq 1\}$ is independent of $\{X'_{nj}, 1 \leq j \leq n, n \geq 1\}$, by (2.11) with $q = 2$, Lemma 2.1 with $p = 2$, c_r-inequality, Lemma 2.3 and (2.1), we have that

$$H_1$$

$$= \sum_{n=1}^{\infty} n^{\alpha p-2-2\alpha} \ell(n) \mathbb{E}\left(\max_{1\leq k\leq n}\left\|\sum_{j=1}^{k}\{A_{nj}X'_{nj} - \mathbb{E}\left[A_{nj}X'_{nj}|\mathcal{G}_{n,j-1}\right]\}\right\|^2\right)$$

$$= \sum_{n=1}^{\infty} n^{\alpha p-2-2\alpha} \ell(n)$$

$$\mathbb{E}\left[\mathbb{E}\max_{1\leq k\leq n}\left\|\sum_{j=1}^{k}\{a_{nj}X'_{nj} - \mathbb{E}\left[a_{nj}X'_{nj}|\mathcal{G}_{n,j-1}\right]\}\right\|^2 \bigg| A_{ni} = a_{ni}, 1 \leq i \leq n\right]$$

$$\leq C_2 \sum_{n=1}^{\infty} n^{\alpha p-2-2\alpha} \ell(n) \mathbb{E}\left[\sum_{j=1}^{n} \mathbb{E}\|a_{nj}X'_{nj}\|^2 \bigg| A_{ni} = a_{ni}, 1 \leq i \leq n\right]$$

$$= C_2 \sum_{n=1}^{\infty} n^{\alpha p-2-2\alpha} \ell(n) \sum_{j=1}^{n} \mathbb{E}\|A_{nj}X'_{nj}\|^2$$

$$= C_2 \sum_{n=1}^{\infty} n^{\alpha p-2-2\alpha} \ell(n) \sum_{j=1}^{n} \mathbb{E}A_{nj}^2 \mathbb{E}\|X'_{nj}\|^2$$

$$\leq C_3 \sum_{n=1}^{\infty} n^{\alpha p-1-2\alpha}\ell(n)\mathbb{E}\left[\|X\|^2\mathbf{1}_{\{\|X\|\leq n^\alpha\}}\right] + C_3 \sum_{n=1}^{\infty} n^{\alpha p-1}\ell(n)\mathbb{P}\{\|X\| > n^\alpha\}$$

$$=:C_3 H_{11} + C_3 H_{12}. \tag{2.15}$$

By Lemma 2.5 and $\mathbb{E}\left[\|X\|^p\ell\left(\|X\|^{1/\alpha}\right)\right] < \infty$, we see that

$$H_{11} = \sum_{n=1}^{\infty} n^{\alpha p-1-2\alpha}\ell(n)\mathbb{E}\left[\|X\|^2\mathbf{1}_{\{\|X\|\leq n^\alpha\}}\right]$$

$$= \sum_{n=1}^{\infty} n^{\alpha p-1-2\alpha}\ell(n)\sum_{j=1}^{n}\mathbb{E}\left[\|X\|^2\mathbf{1}_{\{(j-1)^\alpha<\|X\|\leq j^\alpha\}}\right]$$

$$= \sum_{j=1}^{\infty}\mathbb{E}\left[\|X\|^2\mathbf{1}_{\{(j-1)^\alpha<\|X\|\leq j^\alpha\}}\right]\sum_{n=j}^{\infty} n^{\alpha p-1-2\alpha}\ell(n)$$

$$\leq C_5 \sum_{j=1}^{\infty}\mathbb{E}\left[\|X\|^p\|X\|^{2-p}\mathbf{1}_{\{(j-1)^\alpha<\|X\|\leq j^\alpha\}}\right]j^{\alpha p-2\alpha}\ell(j)$$

$$\leq C_6\mathbb{E}\left[\|X\|^p\ell\left(\|X\|^{1/\alpha}\right)\right]$$

$$< \infty. \tag{2.16}$$

By (2.13), we have that

$$H_{12} = \sum_{n=1}^{\infty} n^{p\alpha-1}\ell(n)\mathbb{P}\{\|X\| > n^\alpha\}$$

$$\leq \sum_{n=1}^{\infty} n^{p\alpha-1-\alpha}\ell(n)\mathbb{E}\left[\|X\|\mathbf{1}_{\{\|X\|>n^\alpha\}}\right]$$

$$\leq C\mathbb{E}[\|X\|^p\ell(\|X\|^{1/\alpha})]$$

$$< \infty. \tag{2.17}$$

Thus we obtain (2.2) by (2.11) and (2.13)-(2.17).

Finally, we will prove (2.3). In fact, for any $\varepsilon > 0$,

$$\sum_{n=1}^{\infty} n^{p\alpha-2-\alpha}\ell(n)\mathbb{E}\left(\max_{1\leq k\leq n}\left\|\sum_{j=1}^{k} A_{nj}X_{nj}\right\| - \varepsilon n^\alpha\right)^+$$

$$= \sum_{n=1}^{\infty} n^{p\alpha-2-\alpha}\ell(n)\int_0^{\infty}\mathbb{P}\left\{\max_{1\leq k\leq n}\left\|\sum_{j=1}^{k} A_{nj}X_{nj}\right\| - \varepsilon n^\alpha > t\right\}dt$$

$$\geq \sum_{n=1}^{\infty} n^{p\alpha-2-\alpha}\ell(n)\int_0^{\varepsilon n^\alpha}\mathbb{P}\left\{\max_{1\leq k\leq n}\left\|\sum_{j=1}^{k} A_{nj}X_{nj}\right\| - \varepsilon n^\alpha > t\right\}dt$$

$$\geq \varepsilon \sum_{n=1}^{\infty} n^{p\alpha-2} \ell(n) \mathbb{P} \left\{ \max_{1 \leq k \leq n} \left\| \sum_{j=1}^{k} A_{nj} X_{nj} \right\| > 2\varepsilon n^{\alpha} \right\}. \qquad (2.18)$$

Thus (2.2) implies (2.3).   ■

We also obtain Theorem 2.2 by Lemmas 2.1, 2.2, 2.3, 2.4 and 2.5.

Proof of Theorem 2.2. Let

$$\mathcal{G}_{n0} = \{\emptyset, \Omega\},$$

$$\mathcal{G}_{nj} = \sigma(X_{n1},\ X_{n2},\ \cdots,\ X_{nj}),$$

$$X'_{nj} = X_{nj} \mathbf{1}_{\{\|X_{nj}\| \leq n^{\alpha}\}}$$

and

$$X''_{nj} = X_{nj} \mathbf{1}_{\{\|X_{nj}\| > n^{\alpha}\}}$$

for $1 \leq j \leq n$ and $n \geq 1$. It can be known that (2.10) holds. By Lemma 2.2 with $a = n^{\alpha}$, for $q > 1$, (2.11) holds. For $p \geq 2$, we obtain that $q > 2(\alpha p - 1)/(2\alpha - 1) \geq 2$. Consequently, for any $1 \leq s \leq 2$, by Hölder's inequality and (2.4), we see that

$$\sum_{j=1}^{n} \mathbb{E}|A_{nj}|^{s}$$

$$\leq \left(\sum_{j=1}^{n} \mathbb{E}|A_{nj}|^{q} \right)^{s/q} \left(\sum_{j=1}^{n} 1 \right)^{1-s/q}$$

$$= O(n). \qquad (2.19)$$

By (2.11), (2.13) and (2.14), we get $H_2 < \infty$ and $H_3 < \infty$.

Following we will prove that $H_1 < \infty$.

Noting that $\{(a_{nj}X'_{nj} - \mathbb{E}[a_{nj}X'_{nj}|\mathcal{G}_{n,j-1}], \mathcal{G}_{nj})\}_{j=1}^{n}$ is a martingale difference sequence for fixed real numbers $a_{n1}, a_{n2}, \cdots, a_{nn}$. For $p \geq 2$, by Lemma 2.1 with $p = q$, we have that

$$H_1$$

$$= \sum_{n=1}^{\infty} n^{\alpha p - 2 - q\alpha} \ell(n)$$

$$\mathbb{E}\left[\mathbb{E} \max_{1 \leq k \leq n} \left\| \sum_{j=1}^{k} \{a_{nj}X'_{nj} - \mathbb{E}\left[a_{nj}X'_{nj}|\mathcal{G}_{n,j-1}\right]\} \right\|^{q} \middle| A_{ni} = a_{ni}, 1 \leq i \leq n \right]$$

$$\leq C_2 \sum_{n=1}^{\infty} n^{\alpha p - 2 - q\alpha} \ell(n)$$

$$\mathbb{E}\left[\left\|\mathbb{E}\left[\sum_{j=1}^{n} \mathbb{E}\left[(a_{nj}X'_{nj} - \mathbb{E}\left[a_{nj}X'_{nj}|\mathcal{G}_{n,j-1}\right])^2 \middle| \mathcal{G}_{n,j-1}\right]\right\|^{\frac{q}{2}} \middle| A_{ni} = a_{ni}, 1 \leq i \leq n\right]$$

$$+ C_2 \sum_{n=1}^{\infty} n^{\alpha p - 2 - q\alpha} \ell(n)$$

$$\mathbb{E}\left[\sum_{j=1}^{n} \mathbb{E}\left\|a_{nj}X'_{nj} - \mathbb{E}\left[a_{nj}X'_{nj}|\mathcal{G}_{n,j-1}\right]\right\|^q \middle| A_{ni} = a_{ni}, 1 \leq i \leq n\right]$$

$$\leq C_2 \sum_{n=1}^{\infty} n^{\alpha p - 2 - q\alpha} \ell(n)\mathbb{E}\left(\sum_{j=1}^{n} \mathbb{E}\left[\left\|A_{nj}X'_{nj} - \mathbb{E}\left[A_{nj}X'_{nj}|\mathcal{G}_{n,j-1}\right]\right\|^2 \middle| \mathcal{G}_{n,j-1}\right]\right)^{q/2}$$

$$+ C_2 \sum_{n=1}^{\infty} n^{\alpha p - 2 - q\alpha} \ell(n) \sum_{j=1}^{n} \mathbb{E}\left\|A_{nj}X'_{nj} - \mathbb{E}\left[A_{nj}X'_{nj}|\mathcal{G}_{n,j-1}\right]\right\|^q$$

$$=: C_2 H_{11} + C_2 H_{12}. \tag{2.20}$$

Since $\{A_{nj}, 1 \leq j \leq n, n \geq 1\}$ is independent of $\{X'_{nj}, 1 \leq j \leq n, n \geq 1\}$, by c_r-inequality and Jensen's inequality, we have that

$$\mathbb{E}\left[\left\|A_{nj}X'_{nj} - \mathbb{E}[A_{nj}X'_{nj}|\mathcal{G}_{n,j-1}]\right\|^2 \middle| \mathcal{G}_{n,j-1}\right]$$

$$\leq 2\left\{\mathbb{E}\left[A_{nj}^2\|X_{nj}\|^2 \mathbf{1}_{\{\|X_{nj}\| \leq n^\alpha\}} \middle| \mathcal{G}_{n,j-1}\right] + \left\|\mathbb{E}\left[A_{nj}X_{nj}\mathbf{1}_{\{\|X_{nj}\| \leq n^\alpha\}} \middle| \mathcal{G}_{n,j-1}\right]\right\|^2\right\}$$

$$\leq 4\mathbb{E}\left[A_{nj}^2\|X_{nj}\|^2 \mathbf{1}_{\{\|X_{nj}\| \leq n^\alpha\}} \middle| \mathcal{G}_{n,j-1}\right]$$

$$\leq 4\mathbb{E}A_{nj}^2\mathbb{E}\left[\|X_{nj}\|^2 \middle| \mathcal{G}_{n,j-1}\right], \quad \text{a.s.}$$

Using (2.19) and $\mathbb{E}\{\sup_{n,j:1\leq j \leq n, n \geq 1} \mathbb{E}[\|X_{nj}\|^2 |\mathcal{G}_{n,j-1}]\}^{q/2} < \infty$, we see that

$$H_{11} \leq 4^{q/2} \sum_{n=1}^{\infty} n^{\alpha p - 2 - q\alpha} \ell(n)\left(\sum_{j=1}^{n} \mathbb{E}A_{nj}^2\right)^{q/2}$$

$$\mathbb{E}\left\{\sup_{1\leq j \leq n, n \geq 1} \mathbb{E}\left[\|X_{nj}\|^2 \middle| \mathcal{G}_{n,j-1}\right]\right\}^{q/2}$$

$$\leq C_3 \sum_{n=1}^{\infty} n^{\alpha p - 2 - q\alpha + q/2} \ell(n)$$

$$< \infty. \tag{2.21}$$

By c_r-inequality, Lemma 2.3 and (2.4), we obtain that

$$H_{12} \leq C_4 \sum_{n=1}^{\infty} n^{\alpha p-2-q\alpha} \ell(n) \sum_{j=1}^{n} \mathbb{E}|A_{nj}|^q \mathbb{E}\left[\|X_{nj}\|^q \mathbf{1}_{\{\|X_{nj}\| \leq n^\alpha\}}\right]$$

$$\leq C_5 \sum_{n=1}^{\infty} n^{\alpha p-1-q\alpha} \ell(n) \mathbb{E}\left[\|X\|^q \mathbf{1}_{\{\|X\| \leq n^\alpha\}}\right]$$

$$+ C_5 \sum_{n=1}^{\infty} n^{\alpha p-1} \ell(n) \mathbb{P}\{\|X\| > n^\alpha\}$$

$$\leq C_5 \sum_{n=1}^{\infty} n^{\alpha p-1-q\alpha} \ell(n) \mathbb{E}\left[\|X\|^q \mathbf{1}_{\{\|X\| \leq n^\alpha\}}\right]$$

$$+ C_5 \sum_{n=1}^{\infty} n^{\alpha p-1-\alpha} \ell(n) \mathbb{E}\left[\|X\| \mathbf{1}_{\{\|X\| > n^\alpha\}}\right]$$

$$= C_5 H_{11}^* + C_5 H_{12}^*. \tag{2.22}$$

By $\alpha > 1/2$, $p \geq 2$ and $q > 2(\alpha p - 1)/(2\alpha - 1)$, we have that $q > p$. So, we get by Lemma 2.5 and $\mathbb{E}[\|X\|^p \ell(\|X\|^{1/\alpha})] < \infty$ that

$$H_{11}^* = \sum_{n=1}^{\infty} n^{p\alpha-1-q\alpha} \ell(n) \sum_{j=1}^{n} \mathbb{E}\left[\|X\|^q \mathbf{1}_{\{(j-1)^\alpha < \|X\| \leq j^\alpha\}}\right]$$

$$= \sum_{j=1}^{\infty} \mathbb{E}\left[\|X\|^q \mathbf{1}_{\{(j-1)^\alpha < \|X\| \leq j^\alpha\}}\right] \sum_{n=j}^{\infty} n^{p\alpha-1-q\alpha} \ell(n)$$

$$\leq C_6 \sum_{j=1}^{\infty} \mathbb{E}\left[\|X\|^p \|X\|^{q-p} \mathbf{1}_{\{(j-1)^\alpha < \|X\| \leq j^\alpha\}}\right] j^{p\alpha-q\alpha} \ell(j)$$

$$\leq C_6 \mathbb{E}[\|X\|^p \ell(\|X\|^{1/\alpha})]$$

$$< \infty. \tag{2.23}$$

By the proof of (2.13), we obtain that

$$H_{12}^* = \sum_{n=1}^{\infty} n^{p\alpha-1-\alpha} \ell(n) \mathbb{E}\left[\|X\| \mathbf{1}_{\{\|X\| > n^\alpha\}}\right]$$

$$\leq C_7 \mathbb{E}[\|X\|^p \ell(\|X\|^{1/\alpha})]$$

$$< \infty. \tag{2.24}$$

Therefore, by (2.20)-(2.24), we get $H_1 < \infty$. Consequently, (2.2) holds for all $\varepsilon > 0$.

The proof of (2.3) in Theorem 2.2 is the same as its proof in Theorem 2.1. ∎

We prove Theorem 2.4 by Lemmas 2.1, 2.2, 2.3 and 2.4.

Proof of Theorem 2.4. Let

$$\mathcal{G}_{n0} = \{\emptyset, \Omega\},$$

$$\mathcal{G}_{nj} = \sigma(X_{n1}, \ X_{n2}, \ \cdots, \ X_{nj}),$$

$$X'_{nj} = X_{nj}\mathbf{1}_{\{\|X_{nj}\| \leq n^\alpha\}}$$

and

$$X''_{nj} = X_{nj}\mathbf{1}_{\{\|X_{nj}\| > n^\alpha\}}$$

for $1 \leq j \leq n$ and $n \geq 1$. It can be seen that (2.10) holds. Similar to the proof of Theorem 2.1, by Lemma 2.2 with $a = n^\alpha$, we have that

$$\sum_{n=1}^{\infty} n^{-2}\mathbb{E}\left(\max_{1 \leq k \leq n}\left\|\sum_{j=1}^{k} A_{nj}X_{nj}\right\| - \varepsilon n^\alpha\right)^+$$

$$\leq C_1 \sum_{n=1}^{\infty} n^{-2-\alpha}\mathbb{E}\left(\max_{1 \leq k \leq n}\left\|\sum_{j=1}^{k}\left\{A_{nj}X'_{nj} - \mathbb{E}[A_{nj}X'_{nj}|\mathcal{G}_{n,j-1}]\right\}\right\|^2\right)$$

$$+ \sum_{n=1}^{\infty} n^{-2}\mathbb{E}\left(\max_{1 \leq k \leq n}\left\|\sum_{j=1}^{k}\left\{A_{nj}X''_{nj} + \mathbb{E}[A_{nj}X'_{nj}|\mathcal{G}_{n,j-1}]\right\}\right\|\right)$$

$$\leq C_1 \sum_{n=1}^{\infty} n^{-2-\alpha}\mathbb{E}\left(\max_{1 \leq k \leq n}\left\|\sum_{j=1}^{k}\left\{A_{nj}X'_{nj} - \mathbb{E}[A_{nj}X'_{nj}|\mathcal{G}_{n,j-1}]\right\}\right\|^2\right)$$

$$+ \sum_{n=1}^{\infty} n^{-2}\mathbb{E}\left(\max_{1 \leq k \leq n}\left\|\sum_{j=1}^{k} A_{nj}X''_{nj}\right\|\right)$$

$$+ \sum_{n=1}^{\infty} n^{-2}\mathbb{E}\left(\max_{1 \leq k \leq n}\left\|\sum_{j=1}^{k}\mathbb{E}[A_{nj}X'_{nj}|\mathcal{G}_{n,j-1}]\right\|\right)$$

$$:= C_1 J_1 + J_2 + J_3. \tag{2.25}$$

Since $\{A_{nj}, 1 \leq j \leq n, n \geq 1\}$ is independent of $\{X_{nj}, 1 \leq j \leq n, n \geq 1\}$, by Lemma 2.3, (2.12) and $\mathbb{E}(\|X\|\ln^+\|X\|) < \infty$, we see that

$$J_2 \leq \sum_{n=1}^{\infty} n^{-2}\sum_{j=1}^{n}\mathbb{E}|A_{nj}|\mathbb{E}\left[\|X_{nj}\|\mathbf{1}_{\{\|X_{nj}\| > n^\alpha\}}\right]$$

$$\leq C_2 \sum_{n=1}^{\infty} n^{-1}\mathbb{E}\left[\|X\|\mathbf{1}_{\{\|X\| > n^\alpha\}}\right]$$

$$=C_2 \sum_{n=1}^{\infty} n^{-1} \sum_{m=n}^{\infty} \mathbb{E}\left[\|X\| \mathbf{1}_{\{m^\alpha < \|X\| \le (m+1)^\alpha\}}\right]$$

$$=C_2 \sum_{m=1}^{\infty} \mathbb{E}\left[\|X\| \mathbf{1}_{\{m^\alpha < \|X\| \le (m+1)^\alpha\}}\right] \sum_{n=1}^{m} n^{-1}$$

$$\le C_3 \sum_{m=1}^{\infty} \ln^+ m \mathbb{E}\left[\|X\| \mathbf{1}_{\{m^\alpha < \|X\| \le (m+1)^\alpha\}}\right]$$

$$\le C_4 \mathbb{E}(\|X\| \ln^+ \|X\|)$$

$$< \infty. \tag{2.26}$$

Next, we will prove $J_3 < \infty$. Since $\{(X_{nj}, \mathcal{G}_{nj}), 1 \le j \le n, n \ge 1\}$ is a triangular array of $\mathbb{B}$-valued martingale differences and $\{A_{nj}, 1 \le j \le n, n \ge 1\}$ is independent of $\{X_{nj}, 1 \le j \le n, n \ge 1\}$, we have that

$$\mathbb{E}\left[A_{nj} X_{nj} | \mathcal{G}_{n,j-1}\right] = \mathbb{E}A_{nj} \mathbb{E}\left[X_{nj} | \mathcal{G}_{n,j-1}\right] = 0, \quad \text{a.s.,} \ 1 \le j \le n, \ n \ge 1.$$

Consequently, by the proof of (2.13), we obtain that

$$J_3 = \sum_{n=1}^{\infty} n^{-2} \mathbb{E}\left(\max_{1 \le k \le n} \left\|\sum_{j=1}^{k} \mathbb{E}[A_{nj} X_{nj} \mathbf{1}_{\{\|X_{nj}\| \le n^\alpha\}} | \mathcal{G}_{n,j-1}]\right\|\right)$$

$$= \sum_{n=1}^{\infty} n^{-2} \mathbb{E}\left(\max_{1 \le k \le n} \left\|\sum_{j=1}^{k} \mathbb{E}[A_{nj} X_{nj} \mathbf{1}_{\{\|X_{nj}\| > n^\alpha\}} | \mathcal{G}_{n,j-1}]\right\|\right)$$

$$\le \sum_{n=1}^{\infty} n^{-2} \sum_{j=1}^{n} \mathbb{E}|A_{nj}| \mathbb{E}\left[\|X_{nj}\| \mathbf{1}_{\{\|X_{nj}\| > n^\alpha\}}\right]$$

$$\le C_2 \sum_{n=1}^{\infty} n^{-1} \mathbb{E}\left[\|X\| \mathbf{1}_{\{\|X\| > n^\alpha\}}\right]$$

$$\le C_4 \mathbb{E}(\|X\| \ln^+ \|X\|)$$

$$< \infty. \tag{2.27}$$

Finally, we turn to prove $J_1 < \infty$. Since $\{(a_{nj} X'_{nj} - \mathbb{E}[a_{nj} X'_{nj} | \mathcal{G}_{n,j-1}], \mathcal{G}_{nj})\}_{j=1}^{n}$ is a martingale difference sequence for fixed real numbers $a_{n1}, a_{n2}, \cdots, a_{nn}$ and $\{A_{nj}, 1 \le j \le n, n \ge 1\}$ is independent of $\{X'_{nj}, 1 \le j \le n, n \ge 1\}$, by Lemma 2.1 with $p = 2$, c_r-inequality, Lemma 2.3, (2.1) and (2.26), we have that

$$J_1 = \sum_{n=1}^{\infty} n^{-2-\alpha}$$

$$\mathbb{E}\left\{\mathbb{E} \max_{1 \le k \le n} \left\|\sum_{j=1}^{k} \{a_{nj} X'_{nj} - \mathbb{E}[a_{nj} X'_{nj} | \mathcal{G}_{n,j-1}]\}\right\|^2 \middle| A_{ni} = a_{ni}, 1 \le i \le n\right\}$$

$$\leq C_2 \sum_{n=1}^{\infty} n^{-2-\alpha} \mathbb{E} \left[\sum_{j=1}^{n} \mathbb{E} \|a_{nj} X'_{nj}\|^2 \,\Big|\, A_{ni} = a_{ni}, 1 \leq i \leq n \right]$$

$$= C_2 \sum_{n=1}^{\infty} n^{-2-\alpha} \sum_{j=1}^{n} \mathbb{E} \|A_{nj} X'_{nj}\|^2$$

$$= C_2 \sum_{n=1}^{\infty} n^{-2-\alpha} \sum_{j=1}^{n} \mathbb{E} A_{nj}^2 \mathbb{E} \|X'_{nj}\|^2$$

$$\leq C_3 \sum_{n=1}^{\infty} n^{-1-\alpha} \mathbb{E} \left[\|X\|^2 \mathbf{1}_{\{\|X\| \leq n^\alpha\}} \right] + C_3 \sum_{n=1}^{\infty} n^{\alpha-1} \mathbb{P} \left\{ \|X\| > n^\alpha \right\}$$

$$\leq C_3 \sum_{n=1}^{\infty} n^{-1-\alpha} \sum_{j=1}^{n} \mathbb{E} \left[\|X\|^2 \mathbf{1}_{\{(j-1)^\alpha < \|X\| \leq j^\alpha\}} \right] + C_3 \sum_{n=1}^{\infty} n^{-1} \mathbb{E} \left[\|X\| \mathbf{1}_{\{\|X\| > n^\alpha\}} \right]$$

$$\leq C_3 \sum_{j=1}^{\infty} \mathbb{E} \left[\|X\|^2 \mathbf{1}_{\{(j-1)^\alpha < \|X\| \leq j^\alpha\}} \right] \sum_{n=j}^{\infty} n^{-1-\alpha} + C_4 \mathbb{E}(\|X\| \ln^+ \|X\|)$$

$$\leq C_5 \sum_{j=1}^{\infty} \mathbb{E} \left[\|X\|^2 \mathbf{1}_{\{(j-1)^\alpha < \|X\| \leq j^\alpha\}} \right] j^{-\alpha} + C_4 \mathbb{E}(\|X\| \ln^+ \|X\|)$$

$$\leq C_6 \mathbb{E} \|X\| + C_4 \mathbb{E}(\|X\| \ln^+ \|X\|)$$

$$< \infty. \tag{2.28}$$

Thus we obtain (2.8) by (2.25)-(2.28).

The proof of (2.9) is similar to that of (2.3) in Theorem 2.1. ∎

Chapter 3

B.P.R.E.

The full name of the title of Chapter 3 is branching process in a random environment.

As usual, we write $\mathbb{N}^* = \{1,\ 2,\ \cdots\}$, $\mathbb{N} = \{0\} \bigcup \mathbb{N}^*$, $\mathbb{R}_+ = [0, \infty)$, and

$$\mathbb{U} = \bigcup_{n=0}^{\infty} (\mathbb{N}^*)^n$$

for the set of all finite sequences, where $(\mathbb{N}^*)^0 = \{\emptyset\}$ contains the null sequence $\emptyset$. We also write $I = (\mathbb{N}^*)^{\mathbb{N}^*}$ for the set of all infinite sequences. For $u \in \mathbb{U}$, write $|u| = n$ for the length of u, and $u|k = u_1 u_2 \cdots u_k$ $(k \leq n)$ for the curtailment of u after k terms. For $|u| = n$ and $v \in \mathbb{U}$ or I, write uv for the sequence obtained by juxtaposition. We partially order $\mathbb{U}$ or I by writing $u \leq v$ to mean that for some $u' \in \mathbb{U}$, $v = uu'$.

Let $\{N_u : u \in \mathbb{U}\}$ be a family of i.i.d. random variables with values in $\mathbb{N}$, defined on some probability space $(\Omega, \mathcal{F}, \mathbb{P})$. For simplicity, we write N for $N_\emptyset$.

Let $\mathbb{T} = \mathbb{T}(\omega)$ be the usual Galton-Watson tree (1986, [107]) with defining elements $\{N_u : u \in \mathbb{U}\}$: (a) $\emptyset \in \mathbb{T}$; (b) if $ui \in \mathbb{T}$, then $u \in \mathbb{T}$; (c) if $u \in \mathbb{T}$ and $i \in \mathbb{N}^*$, then $ui \in \mathbb{T}$ if and only if $1 \leq i \leq N_u$. Here the null sequence $\emptyset$ (of length 0) represents the initial particle; ui represents the ith child of u; N_u represents the number of offspring of the particle u.

Let $\zeta = (\zeta_0,\ \zeta_1,\ \cdots)$ be a sequence of i.i.d. random variables, taking values in some space Θ, whose realization corresponds to a sequence of probability distributions on $\mathbb{N}$:

$$p(\zeta_n) = \{p_i(\zeta_n) : i \geq 0\}, \text{ where } p_i(\zeta_n) \geq 0,\ \sum_{i=0}^{\infty} p_i(\zeta_n) = 1. \tag{3.1}$$

A branching process $(Z_n)_{n\geq 0}$ in the random environment ζ (B.P.R.E.) is a family of time-inhomogeneous branching processes (see e.g. Athreya and Karlin (1971, [8, 9]), Athreya and Ney (1972, [10])): given the environment ζ, the process $(Z_n)_{n \geq 0}$ acts as a Galton-Watson process in varying environments with offspring distributions $p(\zeta_n)$ for particles in nth generation, $n \geq 0$. Let $\mathbb{T}_n$ be the set of all individuals of generation n, denoted by sequences u of positive integers of length $|u| = n$: as usual, the initial particle is denoted by the empty sequence $\emptyset$ (of length 0); if $u \in \mathbb{T}_n$, then $ui \in \mathbb{T}_{n+1}$ if and only if $1 \leq i \leq X_u$. By definition,

$$Z_0 = 1 \quad \text{and} \quad Z_{n+1} = \sum_{u \in \mathbb{T}_n} X_u \quad \text{for} \quad n \geq 0, \tag{3.2}$$

where conditioned on ζ, $\{X_u : |u| = n\}$ are integer-valued random variables with common distribution $p(\zeta_n)$; all the random variables X_u, indexed by finite sequences of integers u, are conditionally independent of each other. The classical Galton-Watson process corresponds to the case where all ζ_n are the same constant.

Let $(\Gamma, \mathbb{P}_\zeta)$ be the probability space under which the process is defined when the environment ζ is given. Therefore under $\mathbb{P}_\zeta$, the random variables X_u are independent of each other, and have the common law $p(\zeta_n)$ if $|u| = n$. The probability $\mathbb{P}_\zeta$ is usually called quenched law. The total probability space can be formulated as the product space $(\Gamma \times \Theta^{\mathbb{N}}, \mathbb{P})$, where $\mathbb{P} = \mathbb{P}_\zeta \otimes \tau$ in the sense that for all measurable and positive functions g, we have

$$\int g \mathrm{d}\mathbb{P} = \int \int g(\zeta, y) \mathrm{d}\mathbb{P}_\zeta(y) \mathrm{d}\tau(\zeta),$$

where τ is the law of the environment ζ. The total probability $\mathbb{P}$ is called annealed law. The quenched law $\mathbb{P}_\zeta$ may be considered to be the conditional probability of the annealed law $\mathbb{P}$ given ζ. The expectation with respect to $\mathbb{P}_\zeta$ (resp. $\mathbb{P}$) will be denoted by $\mathbb{E}_\zeta$ (resp. $\mathbb{E}$).

For $n \geq 0$, write

$$m_n(p) = m_n(p, \zeta) = \sum_{i=0}^{\infty} i^p p_i(\zeta_n) \quad \text{for } p > 0, \quad m_n = m_n(1), \tag{3.3}$$

$$P_0 = 1 \quad \text{and} \quad P_n = m_0 \cdots m_{n-1} \text{ if } n \geq 1. \tag{3.4}$$

Then $\mathbb{E}_\zeta X_u^p = m_n(p)$ if $|u| = n$, and $\mathbb{E}_\zeta Z_n = P_n$ for each n.

We always assume that $p_0(\zeta_0) = 0$ a.s. and $p_1(\zeta_0) < 1$ a.s.

We only consider the supercritical case where

$$\alpha := \mathbb{E} \log m_0 \in (0, \infty]$$

except the following Theorem 3.1. Let

$$m = e^\alpha.$$

It is well-known that under $\mathbb{P}_\zeta$,

$$W_n = \frac{Z_n}{P_n} \qquad (n \geq 0)$$

forms a nonnegative martingale with respect to the filtration

$$\mathcal{E}_0 = \{\emptyset, \Omega\} \qquad \text{and} \qquad \mathcal{E}_n = \sigma\{\zeta, X_u : |u| < n\} \quad \text{for } n \geq 1.$$

It follows that $(W_n, \mathcal{E}_n)$ is also a martingale under $\mathbb{P}$. Let

$$W = \lim_{n \to \infty} W_n, \tag{3.5}$$

where the limit exists a.s. by the martingale convergence theorem, and $\mathbb{E}W \leq 1$ by Fatou's lemma.

We now consider the branching measure μ_ω in the random environment ζ, defined as follows. For the Galton-Watson tree $\mathbb{T} = \mathbb{T}(\omega)$ of branching process in the random environment ζ, we denote the boundary of $\mathbb{T}$ as

$$\partial \mathbb{T} = \{u \in I : (u|n) \in \mathbb{T} \text{ for all } n \in \mathbb{N}\}.$$

As a subset of I, $\partial \mathbb{T}$ is a metrical and compact topological space with

$$B_u = \{v \in \partial \mathbb{T} : u \leq v\} \; (u \in \mathbb{T})$$

its topological basis, and with

$$d(u, v) = e^{-\min\{|u|, |v|\}}$$

a possible metric. If $u \in \mathbb{T}(\omega)$, we write $\mathbb{T}_u(\omega) = \{v \in U : uv \in \mathbb{T}(\omega)\}$ for the shifted tree of $\mathbb{T}(\omega)$ at u. Let $\mu = \mu_\omega$ be the branching measure in the random environment ζ on $\partial \mathbb{T}$: it is the unique Borel measure such that for all $u \in \mathbb{T}$,

$$\mu(B_u) = \frac{W_u}{P_{|u|}},$$

where

$$W_u = \lim_{n \to \infty} \frac{\#\{v \in \mathbb{T}_u : |v| = n\}}{\prod_{j=|u|}^{n+|u|-1} m_j}$$

($\#\{\cdot\}$ denotes the cardinality of the set $\{\cdot\}$), and

$$\mathbb{P}_\zeta(W_u \in \cdot) = \mathbb{P}_{\theta^{|u|}\zeta}(W \in \cdot).$$

From the definition of W_u, one can easily check that

$$\mu(B_u) = W \lim_{n \to \infty} \frac{\#\{v \in \mathbb{T}_n : u \leq v\}}{\#\{v \in \mathbb{T} : |v| = n\}} \qquad \text{for all } u \in \mathbb{T}(\omega).$$

The branching measure is important in the study of branching processes. In deterministic environment case, it has been studied by many authors: see for example Joffe (1978, [63]), O'Brien (1980, [108]), Hawkes (1981, [45]), Lyons *et al.* (1995, [100]), Liu (1996, [87]; 2000, [90]; 2001, [93]), Liu and Rouault (1996, [95]), Liu and Shieh (1999, [97]), Shieh and Taylor (2002, [115]), Duquesne (2009, [33]), Kinnison and Mörters (2010, [68]).

For each $u \in \partial \mathbb{T}$, let $\underline{d}(\mu, u)$ and $\overline{d}(\mu, u)$ be the lower and upper local dimensions of μ at u:

$$\underline{d}(\mu, u) = \liminf_{n \to \infty} \frac{-\log \mu(B_{u|n})}{n}, \quad \overline{d}(\mu, u) = \limsup_{n \to \infty} \frac{-\log \mu(B_{u|n})}{n}.$$

Let

$$m(n) = \min_{u \in \mathbb{T}_n} \mu(B_u) = \min_{u \in \partial \mathbb{T}} \mu(B_{u|n}) \quad \text{and} \quad M(n) = \max_{u \in \mathbb{T}_n} \mu(B_u) = \max_{u \in \partial \mathbb{T}} \mu(B_{u|n}).$$

3.1   Some preliminary results

For $k \geq 0$, let $\phi_{\zeta_k}(s) = \mathbb{E}_\zeta s^{X_u}$, $|u| = k$, and for $n \geq 1$, let $\phi_{0,n}(s) = \mathbb{E}_\zeta s^{Z_n} = \phi_{\zeta_0} \circ \cdots \circ \phi_{\zeta_{n-1}}(s)$, $s \in [0, 1]$.

The following theorem concerns the convergence rate of $\phi_{0,n}(s) - \phi_{0,n}(0)$ as $n \to \infty$.

Theorem 3.1 (a) *Assume that $p_0(\zeta_0) = 0$ a.s. Then $\phi_{0,n}(s) = O\left(1/P_n\right)$ a.s.*
(b) *Assume that $\phi_{\zeta_k}(k \geq 0)$ are linear-fractional generating functions. If $\alpha > 0$ and $\mathbb{E} \log^+ b_0(1) < \infty$ (where $b_0(1) = \lim_{s \uparrow 1} b_0(s) = \phi_{\zeta_0}''(1)/2\phi_{\zeta_0}'(1)^2$), then $\phi_{0,n}(s) - \phi_{0,n}(0) \sim \frac{1}{P_n} \cdot \frac{1}{\left(\sum_{k=0}^{\infty} \frac{b_k(1)}{P_k}\right)^2} \cdot \frac{s}{1-s}$ as $n \to \infty$.*

Proof. We define $b_k(s)$ by

$$\frac{1}{1 - \phi_{\zeta_k}(s)} = \frac{1}{\phi_{\zeta_k}'(1)} \cdot \frac{1}{1 - s} + b_k(s), \quad s \in [0, 1),$$

and $b_k(1) := \lim_{s \uparrow 1} b_k(s) = \phi_{\zeta_k}''(1)/2\phi_{\zeta_k}'(1)^2$, $k \geq 0$. Set

$$\phi_{k,n}(s) = \phi_{\zeta_k} \circ \cdots \circ \phi_{\zeta_{n-1}}(s), \quad s \in [0, 1], \ 0 \leq k \leq n - 1.$$

Let $S_0 = 0$, $S_k = \log \phi_{\zeta_0}'(1) + \cdots + \log \phi_{\zeta_{k-1}}'(1)$, $k \geq 1$. Then

$$S_k = \log P_k, \ k \geq 0. \tag{3.6}$$

By iterating,

$$\frac{1}{1 - \phi_{0,n}(s)} = \frac{e^{-S_n}}{1 - s} + \sum_{k=0}^{n-1} e^{-S_k} b_k(s).$$

Then

$$\phi_{0,n}(s) - \phi_{0,n}(0) = (1 - \phi_{0,n}(0)) - (1 - \phi_{0,n}(s))$$

$$= \frac{1}{e^{-S_n} + \sum_{k=0}^{n-1} e^{-S_k} b_k(0)} - \frac{1}{\frac{e^{-S_n}}{1-s} + \sum_{k=0}^{n-1} e^{-S_k} b_k(s)}$$

$$= \frac{e^{-S_n} \cdot \frac{s}{1-s} + \sum_{k=0}^{n-1} e^{-S_k}(b_k(s) - b_k(0))}{\left[e^{-S_n} + \sum_{k=0}^{n-1} e^{-S_k} b_k(0)\right]\left[\frac{e^{-S_n}}{1-s} + \sum_{k=0}^{n-1} e^{-S_k} b_k(s)\right]}.$$

For each probability generating function ϕ_{ζ_k}, using truncation methods, there exists a probability generating function $\underline{\phi}_{\zeta_k} = \mathbb{E}_\zeta s^{X_u \wedge M}$ satisfying

$$\underline{\phi}''_{\zeta_k}(1) = \mathbb{E}_\zeta (X_u \wedge M)(X_u \wedge M - 1) < +\infty$$

and

$$\phi_{\zeta_k}(s) = \mathbb{E}_\zeta s^{X_u} \leq \mathbb{E}_\zeta s^{X_u \wedge M} = \underline{\phi}_{\zeta_k}(s)$$

for all $s \in [0,1]$, where $|u| = k$ and M is a constant.

By a result of Agresti (1974, [3]), for any probability generating function ϕ_{ζ_k} with $\phi''_{\zeta_k}(1) < \infty$, we can find a linear-fractional generating function $\overline{\phi}_{\zeta_k}$ satisfying $\overline{\phi}'_{\zeta_k}(1) = \phi'_{\zeta_k}(1)$, $\overline{\phi}''_{\zeta_k}(1) = 2\phi''_{\zeta_k}(1)$ and $\phi_{\zeta_k}(s) \leq \overline{\phi}_{\zeta_k}(s)$ for all $s \in [0,1]$. Therefore, without loss of generality, we only consider the case of linear-fractional generating function.

If ϕ_{ζ_k} $(k \geq 0)$ are linear-fractional generating functions, then $b_k(s) = b_k(1)$ $(k \geq 0)$ for all $s \in [0,1)$ and

$$\phi_{0,n}(s) - \phi_{0,n}(0) = \frac{e^{-S_n} \cdot \frac{s}{1-s}}{\left[e^{-S_n} + \sum_{k=0}^{n-1} e^{-S_k} b_k(1)\right]\left[\frac{e^{-S_n}}{1-s} + \sum_{k=0}^{n-1} e^{-S_k} b_k(1)\right]}$$

$$= \frac{\frac{1}{P_n} \cdot \frac{s}{1-s}}{\left[\frac{1}{P_n} + \sum_{k=0}^{n-1} \frac{b_k(1)}{P_k}\right]\left[\frac{1}{P_n(1-s)} + \sum_{k=0}^{n-1} \frac{b_k(1)}{P_k}\right]} \tag{3.7}$$

as (3.6).

If $\alpha > 0$ and $\mathbb{E}\log^+ b_0(1) < \infty$, then

$$\lim_{n\to\infty} P_n = \infty \quad \text{and} \quad \sum_{k=0}^\infty \frac{b_k(1)}{P_k} < \infty \quad \text{a.s.,}$$

so that

$$\phi_{0,n}(s) - \phi_{0,n}(0) \sim \frac{1}{P_n} \cdot \frac{1}{\left(\sum_{k=0}^\infty \frac{b_k(1)}{P_k}\right)^2} \cdot \frac{s}{1-s}$$

as $n \to \infty$. ∎

Proposition 3.1 *Assume $p_0(\zeta_0) = 0$ a.s. and $m < \infty$. Let ρ, c be two numbers in $(0, 1]$. Then the following assertions holds:*

(a) if $1/m < \rho < 1$, then for each constant $\lambda \in \left(\frac{1}{m\rho}, 1\right)$,

$$\phi_{0,n}(1 - c\rho^n) = O(\lambda^n);$$

(b) if $1/m > \rho$, then

$$\lim_{n \to \infty} \phi_{0,n}(1 - c\rho^n) = 1 \quad a.s.$$

In the deterministic environment case, Proposition 3.1 was proved by Liu (2001, [93], Proposition 1.1).

Proof of Proposition 3.1. (a) From the proof of Theorem 3.1, without loss of generality, we can assume that ϕ_{ζ_k} ($k \geq 0$) are linear-fractional generating functions. Thus by (3.7),

$$\phi_{0,n}(1 - c\rho^n) = \frac{e^{-S_n} \cdot \frac{1-c\rho^n}{c\rho^n}}{\left[e^{-S_n} + \sum_{k=0}^{n-1} e^{-S_k} b_k(1)\right]\left[\frac{e^{-S_n}}{c\rho^n} + \sum_{k=0}^{n-1} e^{-S_k} b_k(1)\right]}$$

$$= \frac{\frac{1-c\rho^n}{cP_n\rho^n}}{\left[\frac{1}{P_n} + \sum_{k=0}^{n-1} \frac{b_k(1)}{P_k}\right]\left[\frac{1}{cP_n\rho^n} + \sum_{k=0}^{n-1} \frac{b_k(1)}{P_k}\right]}.$$

Therefore, when $P_n\rho^n \to \infty$,

$$\phi_{0,n}(1 - c\rho^n) \sim \frac{1}{cP_n\rho^n} \cdot \frac{1}{\left(\sum_{k=0}^{n-1} \frac{b_k(1)}{P_k}\right)^2} = O\left(\frac{1}{P_n\rho^n}\right).$$

When $\frac{1}{m\rho} < \lambda < 1$, there exists $\varepsilon_1 > 0$ such that $\frac{1}{(1-\varepsilon_1)m\rho} < \lambda$. By the ergodic theorem, $\frac{\log P_n}{n} \to \log m$ a.s. Therefore a.s., for n large enough,

$$P_n \geq [(1 - \varepsilon_1)m]^n, \text{ and } \frac{1}{P_n\rho^n} \leq \left[\frac{1}{(1-\varepsilon_1)m\rho}\right]^n \leq \lambda^n.$$

This ends the proof of Part (a).

(b) By Jensen's inequality, we have

$$\phi_{0,n}(1 - c\rho^n) = \mathbb{E}_\zeta e^{Z_n \log(1-c\rho^n)} \geq e^{P_n \log(1-c\rho^n)}.$$

If $1/m > \rho$, then $P_n\rho^n \to 0$. Therefore $P_n \log(1 - c\rho^n) \to 0$, so that

$$\liminf_{n \to \infty} \phi_{0,n}(1 - c\rho^n) \geq 1 \quad a.s.$$

Since $\phi_{0,n}(1 - c\rho^n) \leq 1$ a.s., it follows that

$$\lim_{n \to \infty} \phi_{0,n}(1 - c\rho^n) = 1 \quad a.s.$$

∎

Proposition 3.2 *Let B be a Borel set on the real line and set $A = \{W \in B\}$. Define, for $n \geq 0$,*

$$A_n = \{\exists u \in \mathbb{T}_n, P_n \mu(B_u) \in B\} \quad \text{and} \quad A'_n = \{\forall u \in \mathbb{T}_n, P_n \mu(B_u) \in B\}.$$

Then $A_0 = A'_0 = A$, and for all $n \geq 1$,

$$\mathbb{P}_\zeta(A_n) \leq P_n \mathbb{P}_{\theta^n \zeta}(A) \quad a.s.$$

and

$$\mathbb{P}_\zeta(A'_n) = \phi_{0,n}\left(\mathbb{P}_{\theta^n \zeta}(A)\right) \quad a.s.$$

In deterministic environment case, Proposition 3.2 reduces to Proposition 1.2 of Liu (2001, [93]).

Proof of Proposition 3.2. Obviously $A_0 = A'_0 = A$.

Since

$$1_{A_n} \leq \sum_{u \in \mathbb{T}_n} 1\{P_n \mu(B_u) \in B\} = \sum_{u \in \mathbb{T}_n} 1\{W_u \in B\},$$

we have

$$\mathbb{P}_\zeta(A_n) \leq \mathbb{E}_\zeta \sum_{u \in \mathbb{T}_n} 1\{W_u \in B\}$$

$$= \mathbb{E}_\zeta \mathbb{E}_\zeta \left[\sum_{u \in \mathbb{T}_n} 1\{W_u \in B\} \big| \mathbb{T}_n \right]$$

$$= \mathbb{E}_\zeta \sum_{u \in \mathbb{T}_n} \mathbb{E}_\zeta \left[1\{W_u \in B\} \big| \mathbb{T}_n \right]$$

$$= \mathbb{E}_\zeta \sum_{u \in \mathbb{T}_n} \mathbb{E}_\zeta 1\{W_u \in B\}$$

$$= \mathbb{E}_\zeta \sum_{u \in \mathbb{T}_n} \mathbb{E}_{\theta^n \zeta} 1\{W \in B\}$$

$$= \mathbb{E}_\zeta Z_n \mathbb{E}_{\theta^n \zeta} 1\{W \in B\}$$

$$= P_n \mathbb{E}_{\theta^n \zeta} 1\{W \in B\}$$

$$= P_n \mathbb{P}_{\theta^n \zeta}(A) \quad a.s.$$

Since

$$1_{A'_n} = \prod_{u \in \mathbb{T}_n} 1\{P_n \mu(B_u) \in B\} = \prod_{u \in \mathbb{T}_n} 1\{W_u \in B\},$$

we have

$$\mathbb{P}_\zeta(A'_n) = \mathbb{E}_\zeta \prod_{u \in \mathbb{T}_n} \mathbf{1}\{W_u \in B\}$$

$$= \mathbb{E}_\zeta \mathbb{E}_\zeta \left[\prod_{u \in \mathbb{T}_n} \mathbf{1}\{W_u \in B\} \big| \mathbb{T}_n \right]$$

$$= \mathbb{E}_\zeta \prod_{u \in \mathbb{T}_n} \mathbb{E}_\zeta \left[\mathbf{1}\{W_u \in B\} \big| \mathbb{T}_n \right]$$

$$= \mathbb{E}_\zeta \prod_{u \in \mathbb{T}_n} \mathbb{E}_\zeta \mathbf{1}\{W_u \in B\}$$

$$= \mathbb{E}_\zeta \prod_{u \in \mathbb{T}_n} \mathbb{E}_{\theta^n \zeta} \mathbf{1}\{W \in B\}$$

$$= \mathbb{E}_\zeta \left(\mathbb{E}_{\theta^n \zeta} \mathbf{1}\{W \in B\} \right)^{Z_n}$$

$$= \mathbb{E}_\zeta \left(\mathbb{P}_{\theta^n \zeta}(A) \right)^{Z_n}$$

$$= \phi_{0,n} \left(\mathbb{P}_{\theta^n \zeta}(A) \right) \quad \text{a.s.}$$

∎

3.2 An equivalent of $\log m(n)$

In this section, we prove that a.s. $(\log m(n))/n$ has a constant limit that we determine explicitly.

Let $p_- > 0$ be defined by

$$p_- = -\frac{\mathbb{E} \log p_1(\zeta_0)}{\mathbb{E} \log m_0}. \tag{3.8}$$

So $p_- = +\infty$ if $\mathbb{P}(p_1(\zeta_0) = 0) > 0$.

It is known that if $p_1(\zeta_0) > 0$ a.s. and $p_- < +\infty$, then for any $a > p_-$ and $0 < b < p_-$, there exist constants $c_1(\zeta)$, $c_2(\zeta) > 0$ such that

$$c_1(\zeta) x^a \le \mathbb{P}_\zeta(W < x) \le c_2(\zeta) x^b, \ \forall x > 0 \tag{3.9}$$

(see Theorem 4.8 of Hambly (1992, [41])). Therefore

$$p_- = \liminf_{x \to 0} \frac{\log \mathbb{P}_\zeta(W \le x)}{\log x} \quad \text{a.s.} \tag{3.10}$$

We will use the following stronger assumption.

(H1) For any $a > p_-$ and $0 < b < p_-$, there are constants $c_1, c_2 > 0$ independent of the environment such that for almost all ζ and all $x > 0$ small enough,

$$c_1 x^a \le \mathbb{P}_\zeta(W \le x) \le c_2 x^b. \tag{3.11}$$

Sufficient conditions for $\mathbb{P}_\zeta(W \leq x) \leq c_2 x^b$ can be found in Huang and Liu (2012, [60]): Assume that there exist constants $p > 1$ and $A > A_1 > 1$ such that a.s.

$$A_1 \leq m_0 \quad \text{and} \quad m_0(p) \leq A^p \tag{3.12}$$

(recall that m_0 and $m_0(p)$ were defined in (3.3)). (a) If $\|p_1(\zeta_0)\|_\infty < 1$, then for some constants $b > 0$ and $C > 0$ independent of ζ, we have a.s.,

$$\mathbb{P}_\zeta(W \leq x) \leq C x^b, \quad \forall x > 0; \tag{3.13}$$

(b) if $p_1(\zeta_0) = 0$ a.s., then for each $b > 0$, there is some constant $C > 0$ independent of ζ such that for almost all ζ and $x > 0$ small enough,

$$\mathbb{P}_\zeta(W \leq x) \leq C x^b. \tag{3.14}$$

In the following theorem and in all the chapter, we shall write $1/\infty = 0$ by convention.

Theorem 3.2 *Assume that (H1). With probability 1,*

$$\lim_{n \to \infty} \frac{-\log m(n)}{n} = \left(1 + \frac{1}{p_-}\right) \alpha \quad a.s. \tag{3.15}$$

In deterministic environment case, Theorem 3.2 reduces to Theorem 2.1 of Liu (2001, [93]).

We need two lemmas for the proof of Theorem 3.2.

Lemma 3.1 *If there exist some constants $b > 0$ and $c > 0$ such that $\mathbb{P}_\zeta[W \leq x] \leq cx^b$ a.s. for all $x > 0$ small enough, then for all $\eta > (1 + 1/b)\alpha$,*

$$\mathbb{P}_\zeta[m(n) \geq e^{-n\eta} \quad \text{for all } n \in \mathbb{N} \text{ large enough}] = 1 \quad a.s. \tag{3.16}$$

Proof. Notice that $m(n) < e^{-n\eta}$ if and only if $\mu(B_u) < e^{-n\eta}$ for some $u \in \mathbb{T}_n$. So by Proposition 3.2, we have, for all $n \in \mathbb{N}$,

$$\mathbb{P}_\zeta[m(n) < e^{-n\eta}] \leq P_n \mathbb{P}_{\theta^n \zeta}[W < P_n e^{-n\eta}] \quad \text{a.s.}$$

If $\eta > (1 + 1/b)\alpha$, then there exists $\varepsilon_1 > 0$ such that $\eta > (1 + 1/b)(\alpha + \log(1 + \varepsilon_1))$. By the ergodic theorem, $\frac{\log P_n}{n} \to \log m$ a.s. Therefore there exists a $n_0(\zeta)$ dependent on the environment ζ such that

$$P_n \leq ((1 + \varepsilon_1)m)^n$$

for any $n \geq n_0(\zeta)$. Thus

$$\mathbb{P}_\zeta[m(n) < e^{-n\eta}] \leq ((1 + \varepsilon_1)m)^n \mathbb{P}_{\theta^n \zeta}[W < ((1 + \varepsilon_1)m)^n e^{-n\eta}]$$

for all $n \geq n_0(\zeta)$. By our condition, there is a constant c large enough such that for all $x > 0$,

$$\mathbb{P}_\zeta[W \leq x] \leq cx^b \quad \text{a.s.}$$

Hence by the preceding inequality,

$$\mathbb{P}_\zeta[m(n) < e^{-n\eta}] \leq c \left((1 + \varepsilon_1)^{b+1} e^{(b+1)\alpha - b\eta} \right)^n$$

for all $n \geq n_0(\zeta)$. Therefore

$$\sum_{n=1}^{\infty} \mathbb{P}_\zeta[m(n) < e^{-n\eta}] < \infty \quad \text{a.s.}$$

whenever $\eta > (1 + 1/b)\alpha$, and the desired conclusion follows by Borel-Cantelli's lemma.   ∎

Lemma 3.2 (a) $\mathbb{P}_\zeta[m(n) < e^{-n\alpha} \quad \text{for all } n \in \mathbb{N} \text{ large enough}] = 1 \text{ a.s.};$
(b) *if there exist some constants $b > 0$ and $c > 0$ such that $\mathbb{P}_\zeta[W \leq x] \geq cx^b$ a.s. for all $x > 0$ small enough, then for all $\eta < (1 + 1/b)\alpha$,*

$$\mathbb{P}_\zeta[m(n) < e^{-n\eta} \quad \text{for all } n \in \mathbb{N} \text{ large enough}] = 1 \text{ a.s.} \qquad (3.17)$$

Proof. By Borel-Cantelli's lemma, it suffices to prove that

$$\sum_{n=1}^{\infty} \mathbb{P}_\zeta[m(n) \geq e^{-n\eta}] < \infty \quad \text{a.s.}$$

in each of the following cases: (i) $\eta = \alpha$, (ii) the condition of (b) is satisfied and $\alpha < \eta < (1 + 1/b)\alpha$. Notice that $m(n) \geq e^{-n\eta}$ if and only if $\mu(B_u) \geq e^{-n\eta}$ for all $u \in \mathbb{T}_n$; so by Proposition 3.2, we have, for all $n \in \mathbb{N}$,

$$\mathbb{P}_\zeta[m(n) \geq e^{-n\eta}] = \phi_{0,n} \left(\mathbb{P}_{\theta^n \zeta}(W \geq P_n e^{-n\eta}) \right)$$
$$= \phi_{0,n} \left(1 - \mathbb{P}_{\theta^n \zeta}(W < P_n e^{-n\eta}) \right) \quad \text{a.s.}$$

If $\alpha \leq \eta < (1 + 1/b)\alpha$, then there exists $\varepsilon_1 \in (0, 1 - e^{-\alpha/b})$ such that $\alpha \leq \eta < (1 + 1/b)\alpha + \log(1 - \varepsilon_1)$. As $\frac{\log P_n}{n} \to \log m$ a.s., there exists an integer $n_0(\zeta)$ dependent on the environment ζ such that

$$P_n \geq ((1 - \varepsilon_1)m)^n$$

for any $n \geq n_0(\zeta)$. Thus

$$\mathbb{P}_\zeta[m(n) \geq e^{-n\eta}] \leq \phi_{0,n} \left(1 - \mathbb{P}_{\theta^n \zeta}(W < ((1 - \varepsilon_1)m)^n e^{-n\eta}) \right)$$
$$= \phi_{0,n} \left(1 - \mathbb{P}_{\theta^n \zeta}(W < ((1 - \varepsilon_1)e^{\alpha - \eta})^n) \right)$$

for all $n \geq n_0(\zeta)$. By our condition, there is a constant $c_1 \in (0,1)$ small enough such that for all $x \in (0,1]$,

$$\mathbb{P}_\zeta[W < x] \geq c_1 x^b \quad \text{a.s.}$$

Hence by the preceding inequality,

$$\mathbb{P}_\zeta[m(n) \geq e^{-n\eta}] \leq \phi_{0,n}(1 - c_1 \rho^n)$$

for all $n \geq n_0(\zeta)$, where $\rho = ((1-\varepsilon_1)e^{\alpha-\eta})^b$. Since $\rho = (1-\varepsilon_1)^b > 1/m$ if $\eta = \alpha$, by Proposition 3.1, there are some constants $\lambda_1 \in \left(\frac{1}{m\rho}, 1\right)$, $K_1 \in (0, \infty)$ and an integer $n_3(\zeta)$ dependent on the environment ζ such that for all $n \geq n_3(\zeta)$,

$$\phi_{0,n}(1 - c_1 \rho^n) \leq K_1 \lambda_1^n.$$

Therefore

$$\sum_{n=1}^{\infty} \mathbb{P}_\zeta[m(n) \geq e^{-n\eta}] < \infty \quad \text{a.s.}$$

whenever $\eta = \alpha$. Since $\rho = ((1-\varepsilon_1)e^{\alpha-\eta})^b > 1/m$ if $\alpha < \eta < (1+1/b)\alpha$, by Proposition 3.1, there are some constants $\lambda < 1$, $K \in (0, \infty)$ and an integer $n_3(\zeta)$ dependent on the environment ζ such that for all $n \geq n_3(\zeta)$,

$$\phi_{0,n}(1 - c_1 \rho^n) \leq K \lambda^n.$$

Therefore

$$\sum_{n=1}^{\infty} \mathbb{P}_\zeta[m(n) \geq e^{-n\eta}] < \infty \quad \text{a.s.}$$

whenever $\alpha < \eta < (1+1/b)\alpha$. ∎

Proof of Theorem 3.2. For any $a > p_-$ and $0 < b < p_-$, there are constants $c_1, c_2 > 0$ independent of the environment such that for almost all ζ and all $x > 0$ small enough, (3.11) holds, so that the conclusion follows from Lemmas 3.1 and 3.2. ∎

3.3 An equivalent of $\log M(n)$

In this section we find an equivalent of $\log M(n)$ which is similar to that of $\log m(n)$ obtained in the last section.

We assume that $\alpha < \infty$ in the section.

Let $p_+ \in [1, \infty]$ be defined by

$$p_+ = \sup\{a \geq 1 : \mathbb{E}\log \mathbb{E}_\zeta(\frac{Z_1}{m_0})^a < \infty\}. \tag{3.18}$$

Therefore $p_+ = \infty$ if and only if $\mathbb{E}\log\mathbb{E}_\zeta(\frac{Z_1}{m_0})^a < \infty$ for all $a > 1$. Recall that for all fixed $a > 1$, $\mathbb{E}\log\mathbb{E}_\zeta(\frac{Z_1}{m_0})^a < \infty$ and $\alpha < \infty$ if and only if $0 < \mathbb{E}_\zeta W^a < \infty$ a.s. (cf. Theorem 3.2.1 of Huang [59]). So

$$p_+ = \sup\{a \geq 1 : 0 < \mathbb{E}_\zeta W^a < \infty \text{ a.s.}\}.$$

Consequently by Theorem 3.1 of Ramachandran (1962, [113]),

$$p_+ = \liminf_{x\to\infty} \frac{-\log\mathbb{P}_\zeta(W > x)}{\log x} \quad \text{a.s.} \tag{3.19}$$

We shall sometimes need the following assumption.

(H2) For any $a > p_+$ and $0 < b < p_+$, there are constants c_3, $c_4 > 0$ independent of the environment such that for almost all ζ and all $x > 0$ large enough,

$$c_3 x^{-a} \leq \mathbb{P}_\zeta(W \geq x) \leq c_4 x^{-b}. \tag{3.20}$$

It is known that if $m_0 \geq A_1$ a.s. for some constant $A_1 > 1$ and $\|m_0(p)\|_\infty < \infty$ for some $p > 1$, then $\mathbb{E}_\zeta W^p \leq C$ a.s. for some constant $C > 0$, so that

$$\mathbb{P}_\zeta(W > x) \leq Cx^{-p} \quad \text{for all } x > 0. \tag{3.21}$$

This is a direct consequence of (12) and (14) of Li, Hu and Liu (2011, [83]).

The following result is the counter part of Theorem 3.2. Recall that $1/\infty = 0$ by our convention.

Theorem 3.3 *Assume that (H2). With probability 1,*

$$\lim_{n\to\infty} \frac{-\log M(n)}{n} = \left(1 - \frac{1}{p_+}\right)\alpha \quad \text{a.s.} \tag{3.22}$$

In deterministic environment case, Theorem 3.3 reduces to Theorem 3.1 of Liu (2001, [93]).

For the proof, just as in the proof of Theorem 3.2, we first establish two lemmas.

Lemma 3.3 *If there exist some constants $a > 0$ and $c > 0$ such that $\mathbb{P}_\zeta[W > x] \leq cx^{-a}$ a.s. for all $x > 0$ large enough, then for all $\eta < (1 - 1/a)\alpha$,*

$$\mathbb{P}_\zeta[M(n) \leq e^{-n\eta} \quad \text{for all } n \in \mathbb{N} \text{ large enough}] = 1 \text{ a.s.} \tag{3.23}$$

Proof. Notice that $M(n) > e^{-n\eta}$ if and only if $\mu(B_u) > e^{-n\eta}$ for some $u \in \mathbb{T}_n$. Therefore by Proposition 3.2, for all $n \in \mathbb{N}$,

$$\mathbb{P}_\zeta[M(n) > e^{-n\eta}] \leq P_n \mathbb{P}_{\theta^n\zeta}[W > P_n e^{-n\eta}] \quad \text{a.s.}$$

If $\eta < (1 - 1/a)\alpha$, then there exists $\varepsilon_1 > 0$ such that $\eta < (1 - 1/a)\alpha - \frac{1}{a}\log(1 + \varepsilon_1) + \log(1 - \varepsilon_1)$. By the ergodic theorem, $\frac{\log P_n}{n} \to \log m$ a.s., so there exists a $n_0(\zeta)$ dependent on the environment ζ such that

$$((1 - \varepsilon_1)m)^n \le P_n \le ((1 + \varepsilon_1)m)^n$$

for any $n \ge n_0(\zeta)$. Thus

$$\mathbb{P}_\zeta[M(n) > e^{-n\eta}] \le ((1 + \varepsilon_1)m)^n \mathbb{P}_{\theta^n\zeta}[W > ((1 - \varepsilon_1)m)^n e^{-n\eta}]$$

for all $n \ge n_0(\zeta)$. By the condition we can choose a constant $K > 0$ large enough such that for all $x > 0$,

$$\mathbb{P}_\zeta[W > x] \le Kx^{-a} \quad \text{a.s.}$$

Hence by the preceding inequality,

$$\mathbb{P}_\zeta[M(n) > e^{-n\eta}] \le K\left((1 + \varepsilon_1)(1 - \varepsilon_1)^{-a}e^{(1-a)\alpha + a\eta}\right)^n$$

for all $n \ge n_0(\zeta)$. Therefore

$$\sum_{n=1}^{\infty} \mathbb{P}_\zeta[M(n) < e^{-n\eta}] < \infty \quad \text{a.s.}$$

whenever $\eta < (1 - 1/a)\alpha$, and the desired conclusion follows by Borel-Cantelli's lemma. ∎

Lemma 3.4 (a) $\mathbb{P}_\zeta[M(n) > e^{-n\alpha}$ for all $n \in \mathbb{N}$ large enough$] = 1$ a.s.;
(b) if there exist some constants $a > 0$ and $c > 0$ such that $\mathbb{P}_\zeta[W > x] \ge cx^{-a}$ a.s. for all $x > 0$ large enough, then for all $\eta > (1 - 1/a)\alpha$,

$$\mathbb{P}_\zeta[M(n) > e^{-n\eta} \quad \text{for all } n \in \mathbb{N} \text{ large enough}] = 1 \quad \text{a.s.} \qquad (3.24)$$

(c) if there exist some constants $a > 0$ and $c > 0$ such that $\mathbb{P}_\zeta[W > x] \ge cx^{-a}$ a.s. for a non-bounded set of values of $x > 0$, then for all $\eta > (1 - 1/a)\alpha$,

$$\mathbb{P}_\zeta[M(n) > e^{-n\eta} \quad \text{for infinitely many } n \in \mathbb{N}] = 1 \quad \text{a.s.} \qquad (3.25)$$

Proof. Since $M(n) \le e^{-n\eta}$ if and only if $\mu(B_u) \le e^{-n\eta}$ for all $u \in \mathbb{T}_n$, by Proposition 3.2, for all $n \in \mathbb{N}$,

$$\mathbb{P}_\zeta[M(n) \le e^{-n\eta}] = \phi_{0,n}\left(\mathbb{P}_{\theta^n\zeta}(W \le P_n e^{-n\eta})\right) = \phi_{0,n}\left(1 - \mathbb{P}_{\theta^n\zeta}(W > P_n e^{-n\eta})\right) \quad \text{a.s.}$$

Under the condition of (b), there is a constant $c_1 \in (0, 1)$ such that

$$\mathbb{P}_\zeta(W > x) \ge c_1 x^{-a} \quad \text{a.s.}$$

for all $x \geq 1$. If $\alpha \geq \eta > (1 - 1/a)\alpha$, then there exists $\varepsilon_1 \in (0, e^{\alpha/a} - 1)$ such that $\alpha \geq \eta > (1 - 1/a)\alpha + \log(1 + \varepsilon_1)$. As $\frac{\log P_n}{n} \to \log m$ a.s., there exists a $n_0(\zeta)$ dependent on the environment ζ such that

$$P_n \leq ((1 + \varepsilon_1)m)^n$$

for any $n \geq n_0(\zeta)$. Therefore

$$\mathbb{P}_{\theta^n\zeta}(W > P_n e^{-n\eta}) \geq c_1 \left((1 + \varepsilon_1)^{-1} e^{\eta - \alpha}\right)^{na}$$

for any $n \geq n_0(\zeta)$ if $\alpha \geq \eta$, so that by Proposition 3.1,

$$\sum_{n=1}^{\infty} \mathbb{P}_\zeta[M(n) \leq e^{-n\eta}] < \infty \quad \text{a.s.}$$

if either (i) $\eta = \alpha$, or (ii) the condition of (b) is satisfied and $\alpha > \eta > (1 - 1/a)\alpha$. Hence the conclusions in parts (a) and (b) follow from Borel-Cantelli's lemma.

For part (c), notice that if (3.25) holds for some $\eta = \eta_0$, then it also holds for all $\eta > \eta_0$; therefore we need only prove the result for $\alpha > \eta > (1 - 1/a)\alpha$. If $\alpha > \eta > (1 - 1/a)\alpha$, then there exists $\varepsilon_2 > 0$ such that $\alpha > \eta > (1 - 1/a)\alpha + \log(1 + \varepsilon_2)$. As $\frac{\log P_n}{n} \to \log m$ a.s., there exists a $n_0(\zeta)$ dependent on the environment ζ such that

$$P_n \leq ((1 + \varepsilon_2)m)^n$$

for any $n \geq n_0(\zeta)$. Therefore, there are infinitely many $n \in \mathbb{N}^*$ such that

$$\mathbb{P}_{\theta^n\zeta}[W > ((1 + \varepsilon_2)e^{\alpha - \eta})^n] \geq c\left((1 + \varepsilon_2)e^{\alpha - \eta}\right)^{-na} \quad \text{a.s.,}$$

so that by the preceding argument, for all these n,

$$\mathbb{P}_\zeta[M(n) \leq e^{-n\eta}] \leq \phi_{0,n}\left(1 - c\left((1 + \varepsilon_2)e^{\alpha - \eta}\right)^{-na}\right) \quad \text{a.s.}$$

Notice that by Proposition 3.1, the term on the right hand side tends to 0 if $\rho := ((1 + \varepsilon_2)e^{\alpha - \eta})^{-a} > e^{-\alpha} = 1/m$. Therefore for all $\alpha > \eta > (1 - 1/a)\alpha$,

$$\mathbb{P}_\zeta\left(\liminf_{n \to \infty}[M(n) \leq e^{-n\eta}]\right) \leq \liminf_{n \to \infty} \mathbb{P}_\zeta[M(n) \leq e^{-n\eta}]$$

$$\leq \lim_{n \to \infty} \phi_{0,n}\left(1 - c\left((1 + \varepsilon_2)e^{\alpha - \eta}\right)^{-na}\right)$$

$$= 0 \quad \text{a.s.}$$

This implies that (3.25) holds for all $\alpha > \eta > (1 - 1/a)\alpha$. ∎

Proof of Theorem 3.3. Notice that by (H2), for each fixed $0 < b < p_+$, there is some constant $c_4 > 0$ independent of the environment such that for almost all ζ and all $x > 0$ large enough,

$$\mathbb{P}_\zeta(W \geq x) \leq c_4 x^{-b},$$

so that by Lemma 3.3,

$$\liminf_{n\to\infty}(-\log M(n)/n) \ge (1 - 1/p_+)\alpha \quad \text{a.s.}$$

Assume $p_+ < \infty$ and let $a > p_+$ be arbitrarily fixed. Then there is some constant $c_3 > 0$ independent of the environment such that for almost all ζ and all $x > 0$ large enough,

$$\mathbb{P}_\zeta(W \ge x) \ge c_3 x^{-a}.$$

by (H2). So by Lemma 3.4(b),

$$\limsup_{n\to\infty}(-\log M(n)/n) \le (1 - 1/a)\alpha \quad \text{a.s.}$$

The proof is then finished by letting $a \to p_+$.   ∎

3.4   A sufficient condition for no exceptional point

For the local dimensions of μ_ω, we have the following theorem.

Theorem 3.4 *Assume that $p_1(\zeta_0) = 0$ a.s., $m_0 \ge A_1$ a.s. for some constant $A_1 > 1$, and $\|m_0(p)\|_\infty < \infty$ for all $p > 1$. Then for a.s. ζ, $\mathbb{P}_\zeta$-a.s. $\underline{d}(\mu, u) = \overline{d}(\mu, u) = \alpha$ for all $u \in \partial\mathbb{T}$.*

In deterministic environment case, Theorem 3.4 was proved by Liu (2001, [93], Theorem 4.1).

We need two lemmas for the proof of Theorem 3.4.

Lemma 3.5 *Assume that (3.14) and (3.21) hold. With probability 1, for all $u \in \partial\mathbb{T}$, $\left(1 - \frac{1}{p_+}\right)\alpha \le \underline{d}(\mu, u) \le \overline{d}(\mu, u) \le \left(1 + \frac{1}{p_-}\right)\alpha$.*

Proof. By the definitions of $m(n)$ and $M(n)$, for all $u \in \partial\mathbb{T}$,

$$m(n) \le \mu(B_{u|n}) \le M(n),$$

so that

$$\overline{d}(\mu, u) \le \limsup_{n\to\infty}(-\log m(n)/n)$$

and

$$\liminf_{n\to\infty}(-\log M(n)/n) \le \underline{d}(\mu, u).$$

Thus the conclusion comes directly from the proofs of Theorems 3.2 and 3.3.   ∎

Proof of Theorem 3.4. Recall that (3.14) holds if $p_1(\zeta_0) = 0$ a.s. and that (3.21) holds if $m_0 \ge A_1$ a.s. for some constant $A_1 > 1$ and $\|m_0(p)\|_\infty < \infty$ for all $p > 1$. The assertion is a direct consequence of Lemma 3.5.   ∎

Chapter 4

M.C.R.E.

The full name of the title of Chapter 4 is multiplicative cascades in a random environment.

As usual, we write $\mathbb{N}^* = \{1,\ 2,\ \cdots\}$, $\mathbb{R}_+ = [0, \infty)$, $\mathbb{R} = (-\infty, \infty)$ and

$$\mathbb{U} = \bigcup_{n=0}^{\infty} (\mathbb{N}^*)^n$$

for the union of all finite sequences, where $(\mathbb{N}^*)^0 = \{\emptyset\}$ contains the null sequence $\emptyset$. We describe the model of Mandelbrot's multiplicative cascades in a random environment (M.C.R.E.) as follows. Let $\zeta = (\zeta_0, \zeta_1, \cdots) = (\zeta_n)_{n \geq 0}$ be a sequence of i.i.d. random variables taking values in some space Θ, so that each realization of ζ_n corresponds to a probability distribution $F_n(\zeta) = F(\zeta_n)$ on $\mathbb{R}_+$. Suppose that when the environment ζ is given, $\{W_u, u \in \mathbb{U}\}$ is a family of totally independent random variables with values in $\mathbb{R}_+$; all the random variables are defined on some probability space $(\Gamma, \mathbb{P}_\zeta)$; for $u \in \mathbb{U}$, each W_{ui} $(1 \leq i \leq r)$ has distribution $F_n(\zeta) = F(\zeta_n)$ if $|u| = n$, where $|u|$ denotes the length of u. For simplicity, we write W_i for $W_{\emptyset i}$, $1 \leq i \leq r$. The total probability space can be formulated as the product space $(\Gamma \times \Theta, \mathbb{P})$, where $\mathbb{P} = \mathbb{P}_\zeta \otimes \tau$ in the sense that for all measurable and positive functions g, we have

$$\int g \mathrm{d}\mathbb{P} = \int \int g(\zeta, y) \mathrm{d}\mathbb{P}_\zeta(y) \mathrm{d}\tau(\zeta),$$

where τ is the law of the environment ζ. The expectation with respect to $\mathbb{P}_\zeta$ (resp. $\mathbb{P}$) will be denoted by $\mathbb{E}_\zeta$ (resp. $\mathbb{E}$).

Suppose that $\mathbb{E}_\zeta W_1 = 1$ almost surely (a.s.) and $\mathbb{P}(W_1 = 1) < 1$.

Let $\mathcal{F}_0$ be the trivial σ-algebra, and for $n > 1$, let $\mathcal{F}_{n-1}$ be the σ-algebra generated by $\{W_{u_1}, \cdots, W_{u_1 \cdots u_{n-1}} : 1 \leq u_1, \cdots, u_{n-1} \leq r\}$. For $r = 2, 3, \cdots,$

let $Z^{(r)}$ be the Mandelbrot's variable in the random environment ζ associated with W_u ($u \in U/\emptyset$) and parameter r:

$$Z^{(r)} := \lim_{n \to \infty} Y_n^{(r)},$$

where

$$Y_n^{(r)} = \sum_{1 \leq u_1, \, \cdots, \, u_n \leq r} \frac{W_{u_1} \cdots W_{u_1 \cdots u_n}}{r^n}.$$

Let $\mathbb{P}_{\theta\zeta}$ be the probability for the shifted environment $\theta\zeta$. It is easily seen that $Z = Z^{(r)}$ satisfies the following distributional equation:

$$Z^{(r)} = \frac{1}{r} \sum_{i=1}^{r} W_i Z_i^{(r)}, \tag{E}$$

where $Z_i^{(r)}$ are non-negative random variables, which can be chosen independent of each other and independent of $\{W_i, 1 \leq i \leq r\}$ under $\mathbb{P}_\zeta$. Z is a non-negative random variable independent of $Z_i^{(r)}$ and independent of $\{W_i, 1 \leq i \leq r\}$ under $\mathbb{P}_\zeta$, $\mathbb{P}_\zeta\{Z_i^{(r)} \in \cdot\} = \mathbb{P}_{\theta\zeta}\{Z^{(r)} \in \cdot\}$. In terms of Laplace transforms $\phi_\zeta^{(r)}(t) = \mathbb{E}_\zeta \exp\{tZ^{(r)}\}$, the equation reads

$$\phi_\zeta^{(r)}(t) = \left[\mathbb{E}_\zeta \phi_{\theta\zeta}^{(r)}(tW_1/r)\right]^r, \quad \text{a.s.,} \quad t \leq 0. \tag{E'}$$

In the deterministic environment case, the model was first introduced by Mandelbrot (1974, [101]) and is referred to as "microcanonique". For one choice of W_1, $Y_n^{(r)}$ represents a stochastic model for turbulence of Yaglom (1974, [102]), and if $0 < \mathbb{P}(W_1 = 1) = 1 - \mathbb{P}(W_1 = 0)$, $r^n Y_n^{(r)}$ is the n-th generation size of a simple birth-death process. For fixed r, the properties of $Z^{(r)}$ and related subjects have been studied by many authors; see, for example, Kahane and Peyrière (1976, [64]), Durrett and Liggett (1983, [34]), Guivarc'h (1990, [38]), Holley and Waymire (1992, [48]). See also Collet and Koukiou (1992, [27]), Liu (1997, [88]; 1998, [89]; 2000, [91]), Menshikov *et al.* (2005, [106]), Barral *et al.* (2010, [13, 14]) for more general results and for related topics.

Let λ be the Lebesgue measure on $[0, 1]$. Fix $r \geq 2$. For every $n \geq 1$, let μ_r^n be the random measure on $[0, 1]$, having on each r-adic interval $A_{u_1 \cdots u_n}^r = [\sum_{k=1}^{n}(u_k - 1)r^{-k}, \sum_{k=1}^{n}(u_k - 1)r^{-k} + r^{-n})$ the density $W_{u_1} \cdots W_{u_1 \cdots u_n}$ with respect to the Lebesgue measure. In other words,

$$\mu_r^n(f) = \int f \mathrm{d}\mu_r^n = \sum_{1 \leq u_1, \, \cdots, \, u_n \leq r} W_{u_1} \cdots W_{u_1 \cdots u_n} \int_{A_{u_1 \cdots u_n}^r} f \mathrm{d}\lambda, \tag{4.1}$$

for each $f \in \mathcal{L}^1([0, 1], \lambda)$. The mass of μ_r^n is $Y_n^{(r)} = \mu_r^n(1)$.

For fixed $r \geq 2$, a.s. the sequence of random measures $\{\mu_r^n, n \geq 1\}$ is weakly convergent, as $n \to \infty$. Let μ_r^∞ be the Borel extension of this weak limit. The random Borel measure μ_r^∞ on $[0, 1]$ is called the Mandelbrot measure for multiplicative cascades in a random environment. The mass of μ_r^∞ is $Z^{(r)} = \mu_r^\infty(1)$.

In the deterministic environment case, this measure and its extensions have been studied by many authors, see, for example, Kahane and Peyrière (1976, [64]), Waymire and Williams (1996, [144]), Barral (1999, [12]), Liu (2000, [91]), Liu, Rio and Rouault (2003, [94]).

Fix $1 \leq k \leq r$. If the weights $W_{u_1} \cdots W_{u_1 \cdots u_n}$ in (4.1) are replaced by $W_{ku_1} \cdots W_{ku_1 \cdots u_n}$, the corresponding measures will be denoted by $\mu_r^n \circ T_k$ ($1 \leq n < \infty$), i.e.,

$$(\mu_r^n \circ T_k)(f) = \int f \mathrm{d}(\mu_r^n \circ T_k) = \sum_{1 \leq u_1, \cdots, u_n \leq r} W_{ku_1} \cdots W_{ku_1 \cdots u_n} \int_{A_{u_1 \cdots u_n}^r} f \mathrm{d}\lambda,$$

$$(4.2)$$

and its weak limit (as $n \to \infty$) by $\mu_r^\infty \circ T_k$. Notice that the measures μ_r^n and μ_r^∞ depend on the marked r-ary tree with marks $W_{u_1 \cdots u_n}$ associated with each node $u_1 \cdots u_n$, while $\mu_r^n \circ T_k$ and $\mu_r^\infty \circ T_k$ depend on its shift at k. T_k may be considered the shift operator to the node k in the space of marked trees. For fixed r and f, the random variables $(\mu_r^\infty \circ T_k)(f)$, $1 \leq k \leq r$, are independent of each other and independent of $\{W_i, 1 \leq i \leq r\}$ under $\mathbb{P}_\varsigma$, and $\mathbb{P}_\varsigma\{(\mu_r^\infty \circ T_k)(f) \in \cdot\} = \mathbb{P}_{\theta\varsigma}\{\mu_r^\infty(f) \in \cdot\}$. For $k = 1, \cdots, r$, let τ_k^r be the operator acting on functions from $[0, 1]$ to $\mathbb{R}$, defined by

$$\tau_k^r f(x) = f\left(\frac{k - 1 + x}{r}\right), \quad x \in [0, 1].$$

Since $t \in A_{u_1 \cdots u_n}^r$ if and only if $r\left(t - \frac{u_1 - 1}{r}\right) \in A_{u_2 \cdots u_n}^r$, we have, for f in $\mathcal{L}^1([0, 1], \lambda)$,

$$\mu_r^n(f) = \sum_{k=1}^r W_k \sum_{1 \leq u_2, \cdots, u_n \leq r} W_{ku_2} \cdots W_{ku_2 \cdots u_n} \int_{A_{u_2 \cdots u_n}^r} \frac{1}{r} f\left(\frac{s + k - 1}{r}\right) \mathrm{d}s,$$

$$(4.3)$$

so that for each $1 \leq n < \infty$,

$$\mu_r^n(f) = \frac{1}{r} \sum_{k=1}^r W_k(\mu_r^{n-1} \circ T_k)(\tau_k^r f), \tag{4.4}$$

with the convention $\mu_r^0 \circ T_k = \lambda$. Taking the limit as $n \to \infty$ in (4.4), we see that a.s. for every $f \in \mathscr{C}([0, 1])$,

$$\mu_r^\infty(f) = \frac{1}{r} \sum_{k=1}^r W_k(\mu_r^\infty \circ T_k)(\tau_k^r f). \tag{4.5}$$

In the deterministic environment case, this equation and its version for masses $Z^{(r)}$,

$$Z^{(r)} = \frac{1}{r} \sum_{k=1}^{r} W_k(Z^{(r)} \circ T_k), \qquad (4.6)$$

have been studied by many authors (cf. 1976, [64]; 1983, [34]; 1990, [38]; 1998, [89]; 2001, [92]). Asymptotic properties of the masses $Z^{(r)}$ as $r \to \infty$, have been studied by some authors, see, for example, Liu and Rouault (2000, [96]), Liu, Rio and Rouault (2003, [94]).

The purpose of the chapter is to give limit theorems for the process $\{Z^{(r)} : r \geq 2\}$ and the sequence of random measures $(\mu_r^n)_r$ as $r \to \infty$.

4.1   Central limit theorem

Theorem 4.1 (*A central limit theorem*). *If* $\mathbb{E}W_1^2 < \infty$, *then as* $r \to \infty$,

$$\frac{\sqrt{r}}{\sqrt{\mathbb{E}_\zeta W_1^2 - 1}}(Z^{(r)} - 1) \text{ converges in law to the normal law } \mathcal{N}(0,1) \text{ under } \mathbb{P}_\zeta.$$

In the deterministic environment case, Theorem 4.1 reduces to Theorem 1.2 of Liu and Rouault (2000, [96]).

4.2   Convergence in $\mathbb{L}^2$

The following result will be used in the next section.

Theorem 4.2 *If* $\mathbb{E}W_1^2 < r < \infty$, *then*

$$\mathbb{E}(Z^{(r)} - 1)^2 = \mathbb{E}(Z^{(r)})^2 - 1 = \frac{\mathbb{E}W_1^2 - 1}{r - \mathbb{E}W_1^2}.$$

In particular,

$$\lim_{r \to \infty} Z^{(r)} = 1 \ \ in \ \mathbb{L}^2.$$

In the deterministic environment case, Theorem 4.2 reduces to Theorem 3.1 of Liu and Rouault (2000, [96]).

The proof of Theorem 4.2 will be based on the following lemmas.

Lemma 4.1 *Let* $r \geq 2$ *be fixed. Assume that* $\mathbb{E}W_1 \log W_1 \in [-\infty, \infty)$. *Then the following assertions are equivalent:*

(a) $\mathbb{E}W_1 \log W_1 < \log r$;
(b) $\mathbb{E}_\zeta Z^{(r)} = 1$ *a.s.*;
(b') $\mathbb{E}Z^{(r)} = 1$;
(c) $\mathbb{P}_\zeta(Z^{(r)} = 0) < 1$ *a.s.*;
(c') $\mathbb{P}(Z^{(r)} = 0) < 1$.

This is a special case of Theorem 7.1 of Biggins and Kyprianou (2004, [16]) or Theorem 2.5 of Kuhlbusch (2004, [72]).

Lemma 4.2 *Let* $r \geq 2$ *be fixed. For* $\alpha > 1$, *the following assertions are equivalent:*
(a) $\mathbb{E}\left(\sum_{i=1}^r W_i\right)^\alpha < \infty$ *and* $\mathbb{E}W_1^\alpha < r^{\alpha-1}$;
(b) $\mathbb{E}\left(\sup_{n \geq 1} Y_n^{(r)}\right)^\alpha < \infty$;
(c) $0 < \mathbb{E}(Z^{(r)})^\alpha < \infty$.

This is given by Theorem 2.2.2 of Liang (2010, [86]).

Proof of Theorem 4.2. Since the function $f(s) = \log \mathbb{E}W_1^s$ is convex, we have $f(2) - f(1) \geq f'(1)$, which gives $\mathbb{E}W_1 \log W_1 \leq \log \mathbb{E}W_1^2$. Therefore, the condition $\mathbb{E}W_1^2 < r < \infty$ implies $\mathbb{E}W_1 \log W_1 < \log r$, so that by Lemmas 4.1 and 4.2, $\mathbb{E}Z^{(r)} = 1$ and $\mathbb{E}(Z^{(r)})^2 < \infty$. By equation (E), we have,

$$(Z^{(r)})^2 = \frac{1}{r^2}\left(\sum_{i=1}^r W_i Z_i^{(r)}\right)^2 = \frac{1}{r^2}\left[\sum_{i=1}^r W_i^2(Z_i^{(r)})^2 + \sum_{1 \leq i,j \leq r, i \neq j} W_i W_j Z_i^{(r)} Z_j^{(r)}\right],$$

$$\mathbb{E}(Z^{(r)})^2 = \frac{1}{r^2}\left[r\mathbb{E}\mathbb{E}_\zeta W_1^2 \mathbb{E}_{\theta\zeta}(Z^{(r)})^2 + r(r-1)\mathbb{E}(\mathbb{E}_\zeta W_1)^2(\mathbb{E}_{\theta\zeta}Z^{(r)})^2\right]$$

$$= \frac{1}{r}\mathbb{E}\mathbb{E}_\zeta W_1^2 \mathbb{E}_{\theta\zeta}(Z^{(r)})^2 + \frac{r-1}{r}$$

$$= \frac{1}{r}\mathbb{E}W_1^2 \mathbb{E}(Z^{(r)})^2 + \frac{r-1}{r}.$$

So $\mathbb{E}(Z^{(r)})^2 = (r-1)/(r - \mathbb{E}W_1^2)$. Since $\mathbb{E}(Z^{(r)} - 1)^2 = \mathbb{E}(Z^{(r)})^2 - 1$, the desired conclusion holds. ∎

4.3 Proof of Theorem 4.1

In this section, we prove Theorem 4.1.

Proof of Theorem 4.1. Let $r_0 = \mathbb{E}W_1^2$. By the proof of Theorem 4.2, for $r \in (r_0, \infty)$, we have $\mathbb{E}W_1 \log W_1 < \log r$, so that by Lemmas 4.1 and 4.2, for $r \in [r_0, \infty)$, we see that

$$\mathbb{E}_\zeta Z^{(r)} = 1 \quad \text{a.s.},$$

$$\mathbb{E}(Z^{(r)})^2 < \infty$$

and

$$\mathbb{E}_\zeta(Z^{(r)})^2 = \frac{1}{r}\mathbb{E}_\zeta W_1^2 \mathbb{E}_{\theta\zeta}(Z^{(r)})^2 + \frac{r-1}{r} \quad \text{a.s.} \tag{4.7}$$

By equation (E),

$$rZ^{(r)} - r = \sum_{i=1}^{r}(W_i Z_i^{(r)} - 1).$$

Let $S_r = \sum_{i=1}^{r}(W_i Z_i^{(r)} - 1)$ $(r \geq r_0)$ and let $s_r \geq 0$ be defined by

$$s_r^2 = \sum_{i=1}^{r}\mathbb{E}_\zeta(W_i Z_i^{(r)} - 1)^2.$$

We notice that $W_i Z_i^{(r)} - 1$ are totally i.i.d. random variables under $\mathbb{P}_\zeta$ with

$$\mathbb{E}_\zeta[W_i Z_i^{(r)} - 1] = \mathbb{E}_\zeta(W_i Z_i^{(r)}) - 1 = \mathbb{E}_\zeta W_1 \mathbb{E}_{\theta\zeta} Z^{(r)} - 1 = 0 \quad \text{a.s. for} \ r \in [r_0, \infty),$$

and that

$$s_r^2 = r\mathbb{E}_\zeta(W_1 Z_1^{(r)} - 1)^2 = r\left[\mathbb{E}_\zeta W_1^2 \mathbb{E}_\zeta(Z_1^{(r)})^2 - 1\right] = r\left[\mathbb{E}_\zeta W_1^2 \mathbb{E}_{\theta\zeta}(Z^{(r)})^2 - 1\right] \quad \text{a.s.}$$

for $r \in [r_0, \infty)$. We shall verify Lindeberg's condition for the sequence $\{S_r : r \geq r_0\}$. For all $\varepsilon > 0$ and $r \in [r_0, \infty)$, we have

$$\sum_{k=1}^{r}\frac{1}{s_r^2}\int_{\{|W_k Z_k^{(r)}-1|\geq\varepsilon s_r\}}\left[W_k Z_k^{(r)} - 1\right]^2 d\mathbb{P}_\zeta$$

$$=\frac{r}{s_r^2}\int_{\{|W_1 Z_1^{(r)}-1|\geq\varepsilon s_r\}}\left[W_1 Z_1^{(r)} - 1\right]^2 d\mathbb{P}_\zeta$$

$$=\frac{1}{\mathbb{E}_\zeta W_1^2 \mathbb{E}_{\theta\zeta}(Z^{(r)})^2 - 1}\int_{A_r}\left[W_1 Z_1^{(r)} - 1\right]^2 d\mathbb{P}_\zeta$$

$$=\frac{\mathbb{E}_\zeta[W_1 Z_1^{(r)} - 1]^2 \mathbf{1}_{\{A_r\}}}{\mathbb{E}_\zeta W_1^2 \mathbb{E}_{\theta\zeta}(Z^{(r)})^2 - 1}, \tag{4.8}$$

where $A_r = \{|W_1 Z_1^{(r)} - 1| \geq \varepsilon\sqrt{r[\mathbb{E}_\zeta W_1^2 \mathbb{E}_{\theta\zeta}(Z^{(r)})^2 - 1]}\}$. Notice that for $r \in [r_0, \infty)$,

$$\left[W_1 Z_1^{(r)} - 1\right]^2 = W_1^2\left[(Z_1^{(r)})^2 - 1\right] - 2W_1\left[Z_1^{(r)} - 1\right] + (W_1 - 1)^2. \tag{4.9}$$

$$\mathbb{E}\left(W_1^2|(Z_1^{(r)})^2 - 1|\right) = \mathbb{E}\left(\mathbb{E}_\zeta W_1^2 \mathbb{E}_{\theta\zeta}|(Z^{(r)})^2 - 1|\right) = \mathbb{E}W_1^2\mathbb{E}|(Z^{(r)})^2 - 1| \to 0. \tag{4.10}$$

$$\mathbb{E}\big| - 2W_1\big[Z_1^{(r)} - 1\big]\big| = 2\mathbb{E}\left(\mathbb{E}_\zeta W_1 \mathbb{E}_{\theta\zeta}\big|Z^{(r)} - 1\big|\right)$$
$$= 2\mathbb{E}W_1 \mathbb{E}\big|Z^{(r)} - 1\big|$$
$$= 2\mathbb{E}\big|Z^{(r)} - 1\big|$$
$$\to 0. \tag{4.11}$$

Let $\{r'\}$ be any subsequence of $\{r\}$. Notice that from (4.10), we can choose a subsequence $\{r''\}$ of $\{r'\}$ with $r'' \to \infty$ for which

$$\mathbb{E}_\zeta \left(W_1^2\big|(Z_1^{(r'')})^2 - 1\big|\right) \to 0 \quad \text{a.s.} \tag{4.12}$$

Similarly, we also have that

$$\mathbb{E}_\zeta\big| - 2W_1\big[Z_1^{(r'')} - 1\big]\big| \to 0 \quad \text{a.s.} \tag{4.13}$$

By Markov's inequality, we have

$$\mathbb{E}_\zeta \mathbf{1}_{\{A_r\}} = \mathbb{P}_\zeta\{A_r\} \le \frac{\mathbb{E}_\zeta\big[W_1 Z_1^{(r)} - 1\big]^2}{\varepsilon^2 r \big[\mathbb{E}_\zeta W_1^2 \mathbb{E}_{\theta\zeta}(Z^{(r)})^2 - 1\big]} = \frac{1}{\varepsilon^2 r} \to 0 \quad \text{a.s.}$$

Thus

$$\mathbf{1}_{\{A_r\}} \to 0 \quad \text{in probability under } \mathbb{P}_\zeta.$$

Therefore by the dominated convergence theorem, we see that

$$\mathbb{E}_\zeta(W_1 - 1)^2 \mathbf{1}_{\{A_r\}} \to 0 \quad \text{a.s.} \tag{4.14}$$

By (4.8), (4.9), (4.12), (4.13) and (4.14), we have

$$\lim_{r'' \to \infty} \sum_{k=1}^{r''} \frac{1}{s_{r''}^2} \int_{\{|W_k Z_k^{(r'')} - 1| \ge \varepsilon s_{r''}\}} \big[W_k Z_k^{(r'')} - 1\big]^2 d\mathbb{P}_\zeta$$
$$= \lim_{r'' \to \infty} \frac{\mathbb{E}_\zeta\big[W_1 Z_1^{(r'')} - 1\big]^2 \mathbf{1}_{\{A_{r''}\}}}{\mathbb{E}_\zeta W_1^2 \mathbb{E}_{\theta\zeta}(Z^{(r'')})^2 - 1}$$
$$= 0.$$

So by Lindeberg's theorem, $S_{r''}/s_{r''}$ converges in law to the normal law $\mathcal{N}(0,1)$ under $\mathbb{P}_\zeta$. Since

$$\frac{s_{r''}^2}{r''(\mathbb{E}_\zeta W_1^2 - 1)} = \frac{\mathbb{E}_\zeta W_1^2 \mathbb{E}_{\theta\zeta}(Z^{(r'')})^2 - 1}{\mathbb{E}_\zeta W_1^2 - 1} \to 1 \quad \text{a.s. as } r'' \to \infty$$

by Theorem 4.2, this implies that, as $r'' \to \infty$,

$$\frac{\sqrt{r''}}{\sqrt{\mathbb{E}_\zeta W_1^2 - 1}}(Z^{(r'')} - 1) = \frac{S_{r''}}{s_{r''}} \cdot \frac{s_{r''}}{\sqrt{r''(\mathbb{E}_\zeta W_1^2 - 1)}} \quad \text{converges in law to } \mathcal{N}(0,1)$$

under $\mathbb{P}_\zeta$. Since the limit is independent of the subsequence taken, as $r \to \infty$,

$$\frac{\sqrt{r}}{\sqrt{\mathbb{E}_\zeta W_1^2 - 1}}(Z^{(r)}-1) = \frac{S_r}{s_r} \cdot \frac{s_r}{\sqrt{r(\mathbb{E}_\zeta W_1^2 - 1)}} \quad \text{converges in law to } \mathcal{N}(0,1) \text{ under } \mathbb{P}_\zeta.$$

∎

4.4   The Mandelbrot measures for M.C.R.E.

In this section $r \geq 2$ is fixed unless the contrary is mentioned.

Let $f \in \mathcal{L}^1([0,1], \lambda)$ be fixed. The sequence $\{(\mu_r^n(f), \mathcal{F}_n), n \geq 1\}$ is a martingale. By the martingale convergence theorem, considering the positive and negative parts of f, we see that the limit

$$\mu_r(f) = \lim_{n \to \infty} \mu_r^n(f) \tag{4.15}$$

exists $\mathbb{P}_\zeta$-a.s. Let D be a countable dense subset of $\mathscr{C}([0,1])$ equipped with the supremum norm $\| \cdot \|_\infty$. Then $\mathbb{P}_\zeta$-a.s. (4.15) holds for all $f \in \mathscr{C}([0,1])$ since $|\mu_r^n(f)| \leq \|f\|_\infty \mu_r^n(1)$ and $|\mu_r(f)| \leq \|f\|_\infty \mu_r(1)$. Hence $\mathbb{P}_\zeta$-a.s.

$$\mu_r^\infty(f) = \mu_r(f) \quad \text{for all } f \in \mathscr{C}([0,1]) \tag{4.16}$$

(for any Borel measure μ and any integrable function f, we always write $\mu(f) = \int f d\mu$).

In the deterministic environment case, Kahane and Peyrière [64] proved that the positive martingale $\{\mu_r^n(1)\}_n$ is uniformly integrable if and only if $\mathbb{E}W_1 \log W_1 < \log r$. In that case $\mu_r^n(1) \to \mu_r^\infty(1)$ a.s. and in $\mathbb{L}^1$.

Theorem 4.3 *If $\mathbb{E}W_1 \log W_1 < \log r$, then for each fixed $f \in \mathcal{L}^1([0,1], \lambda)$, we have*

$$\lim_{n \to \infty} \mu_r^n(f) = \mu_r^\infty(f) \quad \text{in } \mathbb{L}^1, \quad \text{and} \quad \mu_r^\infty(f) = \mu_r(f) \quad \mathbb{P}_\zeta\text{-a.s.} \tag{4.17}$$

To prove the $\mathbb{L}^1$ convergence, we need the following lemma.

Lemma 4.3 *If $\mathbb{E}W_1 \log W_1 < \log r$, then for each fixed f in $\mathcal{L}^1([0,1], \lambda)$, we have*

$$\mathbb{E}_\zeta \mu_r(f) = \mathbb{E}_\zeta \mu_r^\infty(f) = \lambda(f) \quad \text{a.s.} \tag{4.18}$$

Proof of Lemma 4.3. (a) We first prove that $\mathbb{E}_\zeta \mu_r(f) = \lambda(f)$ a.s. Clearly, for each $1 \leq n < \infty$,

$$\mathbb{E}_\zeta \mu_r^n(f) = \lambda(f) \quad \text{a.s.} \tag{4.19}$$

We assume for the moment that $f \in \mathscr{L}^\infty([0,1], \lambda)$. Since $\mathbb{E}W_1 \log W_1 < \log r$, $\mu_r^n(1) \to \mu_r(1)$ in $\mathbb{L}^1$ by Sheffé's theorem, Lemma 4.1 and (4.15) with $f = 1$. Therefore $\{\mu_r^n(1)\}_n$ is uniformly integrable. As $|\mu_r^n(f)| \leq \|f\|_\infty \mu_r^n(1)$, this implies that $\{\mu_r^n(f)\}_n$ is also uniformly integrable, so that

$$\mu_r^n(f) \to \mu_r(f) \quad \text{in } \mathbb{L}^1 \tag{4.20}$$

by (4.15). Letting $n \to \infty$ in (4.19), we see that $\mathbb{E}_\zeta \mu_r(f) = \lambda(f)$ a.s.

Assume only now $f \in \mathscr{L}^1([0,1], \lambda)$. Fatou's lemma and (4.19) yield $\mathbb{E}_\zeta \mu_r(f) \leq \lambda(f)$ a.s. for $f \geq 0$. Therefore the functional $f \mapsto \mathbb{E}_\zeta \mu_r(f)$ is 1-Lipschitz on $\mathscr{L}^1([0,1], \lambda)$. On $\mathscr{L}^\infty([0,1], \lambda)$, it coincides with the continuous functional $f \mapsto \lambda(f)$. By the density of $\mathscr{L}^\infty([0,1], \lambda)$ in $\mathscr{L}^1([0,1], \lambda)$, this implies that $\mathbb{E}_\zeta \mu_r(f) = \lambda(f)$ a.s. for all $f \in \mathscr{L}^1([0,1], \lambda)$.

(b) We then prove that $\mathbb{E}_\zeta \mu_r^\infty(f) = \lambda(f)$ a.s. Set $\overline{\mu}_r^\infty(A) = \mathbb{E}_\zeta \mu_r^\infty(A)$ for $A \in B$ (recall that B is the Borel σ-field on $[0,1]$). The set function $\overline{\mu}_r^\infty$ is well defined by using the proof of Lemma 2.2 of Liu, Rio and Rouault (2003, [94]). The σ-additivity of μ_r^∞ implies that of $\overline{\mu}_r^\infty$. Therefore $\overline{\mu}_r^\infty$ is a Borel measure on $[0,1]$. For $f \in \mathscr{C}([0,1])$, we have

$$\overline{\mu}_r^\infty(f) = \mathbb{E}_\zeta \mu_r^\infty(f) = \mathbb{E}_\zeta \mu_r(f) = \lambda(f) \quad \text{a.s.}$$

Therefore the measure $\overline{\mu}_r^\infty$ and λ coincide, so that $\mathbb{E}_\zeta \mu_r^\infty(f) = \lambda(f)$ a.s. for all $f \in \mathscr{L}^1([0,1], \lambda)$. ∎

Proof of Theorem 4.3. Fix $f \in \mathscr{L}^1([0,1], \lambda)$. Let $\varepsilon > 0$ be arbitrarily fixed, and take $g \in \mathscr{C}([0,1])$ such that $\lambda(|f - g|) < \varepsilon$. By the triangle inequality and Lemma 4.3,

$$\begin{aligned}
\mathbb{E}_\zeta |\mu_r^n(f) - \mu_r^\infty(f)| &\leq \mathbb{E}_\zeta |\mu_r^n(f-g)| + \mathbb{E}_\zeta |\mu_r^n(g) - \mu_r^\infty(g)| + \mathbb{E}_\zeta |\mu_r^\infty(g-f)| \\
&\leq 2\lambda(|f-g|) + \mathbb{E}_\zeta |\mu_r^n(g) - \mu_r^\infty(g)|.
\end{aligned} \tag{4.21}$$

Because $g \in \mathscr{C}([0,1])$, we have $\lim_{n\to\infty} \mu_r^n(g) = \mu_r^\infty(g) = \mu_r(g)$ in $\mathbb{L}^1$ (cf. (4.20)). Therefore letting $n \to \infty$ in (4.21), we see that

$$\limsup_{n\to\infty} \mathbb{E}_\zeta |\mu_r^n(f) - \mu_r^\infty(f)| \leq 2\varepsilon,$$

so that $\lim_{n\to\infty} \mu_r^n(f) = \mu_r^\infty(f)$ in $\mathbb{L}^1$. Since $\lim_{n\to\infty} \mu_r^n(f) = \mu_r(f)$ $\mathbb{P}_\zeta$-a.s., it follows that $\mu_r^\infty(f) = \mu_r(f)$ $\mathbb{P}_\zeta$-a.s. ∎

Lemma 4.4 (*Proposition 3.1 in Liu, Rio and Rouault (2003, [94])*) *Fix $n \geq 1$ and let $U^1, U^2, \cdots, U^n$ be independent and integrable random variables. Let $(U_{i_1 \cdots i_n}^n)$ be a family of independent random variables indexed by $(n, i_1, \cdots, i_n)$, such that for every n, $U_{i_1 \cdots i_n}^n$ has the same distribution as U^n.*
(a) For $r \geq 1$, set

$$S_r^n = r^{-n} \sum_{1 \leq i_1, \cdots, i_n \leq r} U_{i_1}^1 \cdots U_{i_1 \cdots i_n}^n, \tag{4.22}$$

and let H_r^n be the σ-field generated by $\{S_k^n, k \geq r\}$. Then $\{(S_r^n, H_r^n)\}_{r\geq1}$ is a reverse martingale, and $\lim_{r\to\infty} S_r^n = \mathbb{E}U^1\mathbb{E}U^2 \cdots \mathbb{E}U^n$ a.s. and in $\mathbb{L}^1$.

(b) Assume additionally $\mathbb{E}U^n = 0$. If $\mathbf{a} = \{a_{i_1\cdots i_n}^r, 1 \leq i_1, \cdots, i_n \leq r, r \geq 1\}$ is a family of real numbers such that $\|\mathbf{a}\|_\infty = \sup_{r\geq1} \max_{1\leq i_1, \cdots, i_n\leq r} |a_{i_1\cdots i_n}^r| < \infty$, then as $r \to \infty$,

$$\Gamma_r(\mathbf{a}) := r^{-n} \sum_{1\leq i_1, \cdots, i_n\leq r} U_{i_1}^1 \cdots U_{i_1\cdots i_n}^n a_{i_1\cdots i_n}^r \to 0 \quad \text{a.s. and in } \mathbb{L}^1. \quad (4.23)$$

Lemma 4.5 (*Lemma 3.2 in Liu, Rio and Rouault (2003, [94])*) *Assume that the conditions of Lemma 4.4(b) are satisfied. For $M > 0$, let $\overline{U}_{i_1\cdots i_k}^k := (-M \vee U_{i_1\cdots i_k}^k) \wedge M$. Set $\widetilde{U}_{i_1\cdots i_k}^k := \overline{U}_{i_1\cdots i_k}^k - \mathbb{E}\overline{U}_{i_1\cdots i_k}^k$ and*

$$\Gamma_r^M(\mathbf{a}) := r^{-n} \sum_{1\leq i_1, \cdots, i_n\leq r} \overline{U}_{i_1}^1 \cdots \overline{U}_{i_1\cdots i_{n-1}}^{n-1} \widetilde{U}_{i_1\cdots i_n}^n a_{i_1\cdots i_n}^r. \quad (4.24)$$

Then

$$\lim_{M\to\infty} \limsup_{r\geq1} \sup_{\mathbf{a}:\|\mathbf{a}\|_\infty\leq1} |\Gamma_r(\mathbf{a}) - \Gamma_r^M(\mathbf{a})| = 0 \quad \text{a.s.} \quad (4.25)$$

Lemma 4.6 (*Proposition 3.4 in Liu, Rio and Rouault (2003, [94])*) *Let $\{U_{nk}, n \geq 1, 1 \leq k \leq r_n\}$ be a triangular array of row-wise independent, integrable and centered real random variables such that $\lim_{n\to\infty} r_n = \infty$. If the family $\{U_{nk}, n \geq 1, 1 \leq k \leq r_n\}$ is uniformly integrable, then as $n \to \infty$,*

$$U_n = \frac{1}{r_n} \sum_{k=1}^{r_n} U_{nk} \to 0 \quad \text{in } \mathbb{L}^1. \quad (4.26)$$

For $n \leq \infty$ and some subset G of $\mathscr{L}^1([0,1], \lambda)$, we shall study a.s. and $\mathbb{L}^1$ convergence of

$$\|\mu_r^n - \lambda\|_G := \sup_{f\in G} |\mu_r^n(f) - \lambda(f)| \quad (4.27)$$

as $r \to \infty$. In order to obtain uniform convergence results for finite n, we need finiteness of metric entropy in $\mathscr{L}^1([0,1], \lambda)$.

Definition 4.1 (*Definition 3.6 in Liu, Rio and Rouault (2003, [94])*) *Let (V, d) be an arbitrary semi-metric space and T be a subset of V. The covering number $N(\varepsilon, T, d)$ is the minimal number of balls of radius ε needed to cover T. The entropy number is $H(\varepsilon, T, d) = \log N(\varepsilon, T, d)$. The subset T is said to be totally bounded in (V, d) if $N(\varepsilon, T, d)$ is finite for all $\varepsilon > 0$.*

Definition 4.2 (*Definition 3.7 in Liu, Rio and Rouault (2003, [94])*) *For $f, g \in \mathscr{L}^1([0,1], \lambda)$ such that $f \leq g$, the bracket $[f, g]$ is the set of all $h \in \mathscr{L}^1([0,1], \lambda)$ such that $f \leq h \leq g$. It is called an ε-bracket if $\lambda(g - f) \leq \varepsilon$. The class G is said to be totally bounded with brackets in $\mathscr{L}^1([0,1], \lambda)$ if it can be covered by a finite number of ε-brackets, for all $\varepsilon > 0$.*

Theorem 4.4 *Let* $1 \leq n < \infty$ *be fixed.*
(a) $\lim_{r \to \infty} Y_n^{(r)} = 1$ $\mathbb{P}_\zeta$*-a.s. and in* $\mathbb{L}^1$.
(b) *For* $f \in \mathscr{L}^1([0,1], \lambda)$,

$$\lim_{r \to \infty} \mu_r^n(f) = \lambda(f) \quad in \ \mathbb{L}^1.$$

(c) *If* G *is a class of uniformly bounded functions, totally bounded in* $\mathscr{L}^1([0,1], \lambda)$, *then*

$$\lim_{r \to \infty} \|\mu_r^n - \lambda\|_G = 0 \quad \mathbb{P}_\zeta\text{-a.s. and in } \mathbb{L}^1. \tag{4.28}$$

In the deterministic environment case, Theorem 4.4 reduces to Theorem 3.8 of Liu, Rio and Rouault (2003, [94]).

Proof of Theorem 4.4. Part (a) is a direct consequence of Lemma 4.4(a).

To prove parts (b) and (c), we first remark that for each $f \in \mathscr{L}^\infty([0,1], \lambda)$ and $1 \leq n < \infty$,

$$\lim_{r \to \infty} \left(\mu_r^n(f) - \mu_r^{n-1}(f) \right) = 0 \quad \mathbb{P}_\zeta\text{-a.s. and in } \mathbb{L}^1, \tag{4.29}$$

by applying Lemma 4.4(b) to the decomposition

$$\mu_r^n(f) - \mu_r^{n-1}(f) = \sum_{1 \leq u_1, \cdots, u_n \leq r} W_{u_1} \cdots W_{u_1 \cdots u_{n-1}} (W_{u_1 \cdots u_n} - 1) \int_{A_{u_1 \cdots u_n}^r} f d\lambda.$$

Since $\mu_r^0 = \lambda$, (4.29) implies that, for each $f \in \mathscr{L}^\infty([0,1], \lambda)$ and $1 \leq n < \infty$,

$$\lim_{r \to \infty} \left(\mu_r^n(f) - \lambda(f) \right) = 0 \quad \mathbb{P}_\zeta\text{-a.s. and in } \mathbb{L}^1. \tag{4.30}$$

By the density of $\mathscr{L}^\infty([0,1], \lambda)$ in $\mathscr{L}^1([0,1], \lambda)$, $\mathbb{E}_\zeta \mu_r^n(f) = \lambda(f)$ a.s. for each $1 \leq n < \infty$ and the inequality

$$|\mu_r^n(f) - \lambda(f)| \leq \mu_r^n(|f - g|) + |\mu_r^n(g) - \lambda(g)| + \lambda(|g - f|), \tag{4.31}$$

we see that the $\mathbb{L}^1$ convergence in (4.30) still holds for every f in $\mathscr{L}^1([0,1], \lambda)$, which ends the proof of (b).

For part (c), we assume that G is uniformly bounded by 1 for the sake of simplicity. To prove the a.s. convergence, it is enough to show that for every $n < \infty$,

$$\lim_{r \to \infty} \|\mu_r^n - \mu_r^{n-1}\|_G = 0 \quad \mathbb{P}_\zeta\text{-a.s.} \tag{4.32}$$

From Lemma 4.5, it is sufficient to prove (4.32) when the W_u are bounded by a constant $M \geq 1$. Since G is totally bounded, for every $\varepsilon > 0$ one can find $f_1, \cdots, f_N \in \mathscr{L}^1([0,1], \lambda)$ such that for every $f \in G$ there is some f_i such that $\lambda(|f - f_i|) \leq \varepsilon$. Actually we can choose the functions f_i in $\mathscr{L}^\infty([0,1], \lambda)$ since it is dense in $\mathscr{L}^1([0,1], \lambda)$.

By definition of μ_r^n, we then have $|\mu_r^n(g)| \leq M^n \lambda(|g|)$ for g in $\mathscr{L}^1([0,1], \lambda)$ and $n \geq 0$. Hence, for $f \in G$ and $\lambda(|f - f_i|) \leq \varepsilon$,

$$\left| \left(\mu_r^n(f) - \mu_r^{n-1}(f) \right) - \left(\mu_r^n(f_i) - \mu_r^{n-1}(f_i) \right) \right| \leq 2M^n \varepsilon. \tag{4.33}$$

Now, from (4.30) $\mathbb{P}_\zeta$-a.s. for every $1 \leq i \leq N$,

$$\lim_{r \to \infty} \left(\mu_r^n(f_i) - \mu_r^{n-1}(f_i) \right) = 0. \tag{4.34}$$

Jointly with (4.33) it yields $\mathbb{P}_\zeta$-a.s.

$$\limsup_{r \to \infty} \|\mu_r^n - \mu_r^{n-1}\|_G \leq 2M^n \varepsilon \tag{4.35}$$

for every ε. This gives the $\mathbb{P}_\zeta$-a.s. convergence of part (c).

To get the $\mathbb{L}^1$ convergence, it is enough to prove, for every fixed $n < \infty$, the uniform integrability of $(\|\mu_r^n - \mu_r^{n-1}\|_G)_r$. But this is indeed the case because $\|\mu_r^n - \mu_r^{n-1}\|_G$ is bounded by

$$S_r^n := r^{-n} \sum_{1 \leq u_1, \cdots, u_n \leq r} W_{u_1} \cdots W_{u_1 \cdots u_{n-1}} |W_{u_1 \cdots u_n} - 1| \tag{4.36}$$

which by Lemma 4.4 converges in $\mathbb{L}^1$ and is therefore uniformly integrable. ∎

Theorem 4.5 *Assume* $\mathbb{E}W_1 \log^+ W_1 < \infty$.
(a) $\lim_{r \to \infty} Z^{(r)} = 1$ $\mathbb{P}_\zeta$-*a.s. and in* $\mathbb{L}^1$.
(b) *For* $f \in \mathscr{L}^1([0,1], \lambda)$,

$$\lim_{r \to \infty} \mu_r^\infty(f) = \lambda(f) \text{ in } \mathbb{L}^1.$$

(c) *If* G *is a subset of* $\mathscr{L}^1([0,1], \lambda)$ *such that, for each* $\varepsilon > 0$, *it can be covered by a finite number of* ε-*brackets* $[f_i, g_i]$, *with* f_i *and* g_i *measurable, bounded and* λ-*a.e. continuous, then*

$$\lim_{r \to \infty} \mathbb{E}_\zeta^* \|\mu_r^\infty - \lambda\|_G = 0 \text{ and } \lim_{r \to \infty} \|\mu_r^\infty - \lambda\|_G = 0 \mathbb{P}_\zeta^*\text{-a.s.}, \tag{4.37}$$

where $\mathbb{P}_\zeta^*$ *and* $\mathbb{E}_\zeta^*$ *denote the corresponding outer conditional probability and outer conditional expectation.*

In the deterministic environment case, Theorem 4.5 reduces to Theorem 3.9(a)-(c) of Liu, Rio and Rouault (2003, [94]).

Proof of Theorem 4.5. (a) For $n \leq +\infty$, let $H_n^{(r)}$ be the σ-field generated by $Y_n^{(s)}$, $s \geq r$. By Lemma 4.4(a), for each $n < \infty$, $\{(Y_n^{(r)}, H_n^{(r)})\}_{r \geq 1}$ is a reverse martingale. Thus for every integer $p \geq 1$ and every bounded and continuous function $g : \mathbb{R}^p \to \mathbb{R}$, we have

$$\mathbb{E}_\zeta \left(Y_n^{(r)} g(Y_n^{(r+1)}, \ Y_n^{(r+2)}, \ \cdots, \ Y_n^{(r+p)}) \right)$$

$$= \mathbb{E}_\zeta \left(Y_n^{(r+1)} g(Y_n^{(r+1)}, \ Y_n^{(r+2)}, \ \cdots, \ Y_n^{(r+p)}) \right). \tag{4.38}$$

Let $r_0 \geq 2$ be such that $\mathbb{E}W_1 \log W_1 < \log r_0$. For each fixed $r \geq r_0$, as $n \to \infty$, $Y_n^{(r)} \to Z^{(r)}$ $\mathbb{P}_\zeta$-a.s. and in L^1. Thus using uniform integrability, we may let $n \to \infty$ in (4.38), showing that $\{(Z^{(r)}, H_\infty^{(r)})\}_{r \geq 1}$ is also a reverse martingale. Therefore $Z^{(r)}$ convergence $\mathbb{P}_\zeta$-a.s. and is uniformly integrable. To identify the limit, we will see in (b) below that $Z^{(r)} \to 1$ in L^1, so that the proof of (a) is finished.

(b) We first prove that for each $f \in \mathscr{L}^\infty([0,1], \lambda)$,

$$\lim_{r \to \infty} \mu_r^\infty(f) = \lambda(f) \quad \text{in } L^1. \tag{4.39}$$

By extension of (4.4) to the associated Borel measures we get the decomposition

$$\mu_r^\infty(f) - \lambda(f) = \frac{1}{r} \sum_{k=1}^r \left[W_k(\mu_r^\infty \circ T_k)(\tau_k^r f) - \lambda(\tau_k^r f) \right]. \tag{4.40}$$

Since $\left| W_k(\mu_r^\infty \circ T_k)(\tau_k^r f) - \lambda(\tau_k^r f) \right| \leq c_1 Z^{(r)} \circ T_k + c_2$ (c_1, c_2 are constants), the family $\{W_k(\mu_r^\infty \circ T_k)(\tau_k^r f) - \lambda(\tau_k^r f)\}_{k,r}$ is uniformly integrable, so that Lemma 4.6 gives (4.39). By density of $\mathscr{L}^\infty([0,1], \lambda)$ in $\mathscr{L}^1([0,1], \lambda)$, using Lemma 4.3 and

$$|\mu_r^\infty(f) - \lambda(f)| \leq \mu_r^\infty(|f - g|) + |\mu_r^\infty(g) - \lambda(g)| + \lambda(|g - f|)$$

for $g \in \mathscr{L}^\infty([0,1], \lambda)$, we see that (4.39) holds for $f \in \mathscr{L}^1([0,1], \lambda)$.

(c) Let us first reduce the problem to a simpler one involving only one function. Let $\varepsilon > 0$, and let $\{[f_i, g_i] : 1 \leq i \leq N\}$ be a cover of G by ε-brackets, with f_i and g_i measurable, bounded and λ-a.e. continuous. If $f \in [f_i, g_i]$, then

$$\mu_r^\infty(f) - \lambda(f) \leq \mu_r^\infty(g_i) - \lambda(f_i) = [\mu_r^\infty(g_i) - \lambda(g_i)] + [\lambda(g_i) - \lambda(f_i)]$$

and

$$\mu_r^\infty(f) - \lambda(f) \geq \mu_r^\infty(f_i) - \lambda(g_i) = [\mu_r^\infty(f_i) - \lambda(f_i)] + [\lambda(f_i) - \lambda(g_i)].$$

Therefore

$$\|\mu_r^\infty - \lambda\|_G \leq \max\{|\mu_r^\infty(g_i) - \lambda(g_i)|, |\mu_r^\infty(f_i) - \lambda(f_i)| : 1 \leq i \leq N\} + \varepsilon. \tag{4.41}$$

(c1) To prove the $\mathbb{P}_\zeta^*$-a.s. convergence, it is convenient to introduce the random measures $\widetilde{\mu}_r^n$ defined by

$$\widetilde{\mu}_r^n = \frac{1}{r} \sum_{k=1}^r W_k(Y_{n-1}^{(r)} \circ T_k) \delta_{\frac{k}{r}}, 1 \leq n \leq \infty, \tag{4.42}$$

(recall that by convention $Y_{n-1}^{(r)} \circ T_k = 1$ if $n = 1$, and $= Z^{(r)} \circ T_k$ if $n = \infty$), and to compare it with μ_r^n with the help of (4.4).

Let us first prove that $\mathbb{P}_\zeta$-a.s. for all $t \in [0,1]$,

$$\lim_{r \to \infty} \widetilde{\mu}_r^\infty([0,t]) = t. \tag{4.43}$$

For fixed $t \in (0,1]$ and $1 \le n < \infty$, set

$$^t Y_n^{(r)} := \frac{r}{[rt]} \widetilde{\mu}_r^n([0,t]) = \frac{1}{[rt]} \sum_{u_1=1}^{[rt]} W_{u_1} \sum_{1 \le u_2, \cdots, u_n \le r} \frac{W_{u_1 u_2} \cdots W_{u_1 \cdots u_n}}{r^{n-1}}, \tag{4.44}$$

where $[x]$ is the integer part of x. By Theorem 4.3 and (4.44), if $\mathbb{E} W_1 \log W_1 < \log r$, then as $n \to \infty$, $^t Y_n^{(r)}$ converges $\mathbb{P}_\zeta$-a.s. and in $\mathbb{L}^1$ to

$$^t Y_\infty^{(r)} := \frac{1}{[rt]} \sum_{k=1}^{[rt]} W_k Z^{(r)} \circ T_k. \tag{4.45}$$

For $1 \le n \le \infty$, let $^t H_n^{(r)}$ be the σ-field generated by $\{^t Y_n^{(k)}, k \ge r\}$. Let $r \ge t^{-1}$ be such that $\mathbb{E} W_1 \log W_1 < \log r$. Just like $Y_n^{(r)}$, for each fixed $1 \le n \le \infty$, the sequence $\{^t Y_n^{(r)}\}_{r \ge r_t}$ is a reverse martingale with respect to $\{^t H_n^{(r)}\}_{r \ge t^{-1}}$ (the proof is similar with that of (a)), so that it converges $\mathbb{P}_\zeta$-a.s. and in L^1. To identify the limit of $^t Y_\infty^{(r)}$, we use

$$^t Y_\infty^{(r)} - 1 = \frac{1}{[rt]} \sum_{k=1}^{[rt]} (W_k Z^{(r)} \circ T_k - 1) \tag{4.46}$$

and Lemma 4.6 to conclude that $^t Y_\infty^{(r)} \to 1$ in L^1. Since

$$\widetilde{\mu}_r^\infty([0,t]) = \frac{[rt]}{r} \, {}^t Y_\infty^{(r)},$$

it follows that

$$\lim_{r \to \infty} \widetilde{\mu}_r^\infty([0,t]) = t \quad \mathbb{P}_\zeta\text{-a.s. and in } L^1.$$

By a classical monotonicity argument, this implies (4.43), hence the $\mathbb{P}_\zeta$-a.s. weak convergence of $\widetilde{\mu}_r^\infty$ to λ. To get a similar result for μ_r^∞, observe first that, from (4.5),

$$\mu_r^\infty(f) - \widetilde{\mu}_r^\infty(f) = \frac{1}{r} \sum_{k=1}^r W_k (\mu^\infty \circ T_k)(\tau_k^r f - f(k/r)). \tag{4.47}$$

Since, for $f \in \mathscr{C}([0,1])$,

$$\sup_{x \in [0,1]} |\tau_k f(x) - f_{k,r}| \le \omega_f(r^{-1}),$$

where $\omega_f(h)$ is the maximal oscillation of f on intervals of size h, $h > 0$, we have

$$|\mu_r^\infty(f) - \tilde{\mu}_r^\infty(f)| \le \frac{\omega_f(r^{-1})}{r} \sum_{k=1}^{r} W_k(Z^{(r)} \circ T_k) = \omega_f(r^{-1})Z^{(r)}, \qquad (4.48)$$

where the last equality holds by (4.6). This yields the $\mathbb{P}_\zeta$-a.s. weak convergence of μ_r^∞ to λ. Therefore (cf. [20], p. 163, Proposition 8.12) $\mathbb{P}_\zeta^*$-a.s. for all f measurable, bounded and λ-a.s. continuous,

$$\lim_{r\to\infty} \mu_r^\infty(f) = \lambda(f). \qquad (4.49)$$

Replacing f by f_i, g_i in the above equation and using (4.41), we see that

$$\mathbb{P}_\zeta^*\text{-a.s.} \quad \limsup_{r\to\infty} \|\mu_r^\infty - \lambda\|_G \le \varepsilon,$$

for every $\varepsilon > 0$, which ends the proof of the $\mathbb{P}_\zeta^*$-a.s. convergence.

(c2) Taking $\mathbb{E}_\zeta^*$ in (4.41) and using (b) gives the $\mathbb{L}^1$-convergence. ∎

Chapter 5

D.P.R.E.

The full name of the title of Chapter 5 is directed polymers in a random environment.

Concentration inequalities are very powerful tools in probability theory and have been studied by many authors; see e.g.: Talagrand (1995, [125]; 1996, [126]), Ledoux (1999, [77]) for related concentration inequalities and Wang (2005, [137]) for general functional inequalities.

For notations, as usual, we write $\mathbb{N}^* = \{1, 2, \cdots\}$, $\mathbb{N} = \{0\} \bigcup \mathbb{N}^*$ and $\mathbb{R}^+ = (0, +\infty)$.

Let $d \in \mathbb{N}^*$ and $\omega = (\omega_n)_{n \in \mathbb{N}}$ be the simple random walk on the d-dimensional integer lattice $\mathbb{Z}^d$ starting at 0, defined on a probability space $(\Omega, \mathcal{F}, \mathbb{P})$. Let $\eta = (\eta(n, x))_{(n,x) \in \mathbb{N} \times \mathbb{Z}^d}$ be a sequence of real-valued, non-constant, i.i.d. random variables defined on another probability space $(E, \mathcal{E}, \mathbb{Q})$ (we use the letter E to refer to the environment). We denote by F the common distribution function of the sequence $(\eta(n, x))_{(n,x) \in \mathbb{N} \times \mathbb{Z}^d}$. The path ω represents the directed polymer and the letter η the environment sequence $(\eta(n, x))_{(n,x) \in \mathbb{N} \times \mathbb{Z}^d}$. For any $n > 0$, define the random polymer measure μ_n on the path space $(\Omega, \mathcal{F})$ by

$$\mu_n = \frac{1}{Z_n(\beta)} \exp(\beta H_n(\omega)) \mathbb{P}(d\omega), \tag{5.1}$$

where $\beta > 0$ is the inverse temperature,

$$H_n(\omega) = \sum_{j=1}^{n} \eta(j, \omega_j) \quad \text{and} \quad Z_n(\beta) = \mathbb{P}[\exp(\beta H_n(\omega))]. \tag{5.2}$$

Let $\lambda(\beta) = \ln \mathbb{Q}\left[e^{\beta \eta(0,0)}\right] \leq +\infty$ be the logarithmic moment generating function of $\eta(0,0)$. Let

$$W_n(\beta) = \mathbb{P}\left[\exp\left(\beta \sum_{j=1}^{n} \eta(j,\omega_j) - n\lambda(\beta)\right)\right] \tag{5.3}$$

be the normalized partition function if $\mathbb{Q}[e^{\beta|\eta(0,0)|}] < +\infty$ for fixed $\beta > 0$ (the condition is equivalent to $\lambda(\pm\beta) < +\infty$).

This model first appeared in the physics literature (see Huse and Henley (1985, [46])) for modeling the phase boundary of the Ising model subject to random impurities, the first mathematical study was undertaken by Imbrie and Spencer (1988, [61]) and Bolthausen (1989, [19]). For recent results, see e.g.: Albeverio and Zhou (1996, [5]), Kifer (1997, [67]), Carmona and Hu (2002, [22]; 2004, [23]), Comets *et al.* (2004, [29]), Birkner (2004, [18]), Mejane (2004, [105]), Carmona *et al.* (2006, [21]), Comets and Yoshida (2006, [31]), Comets and Vargas (2006, [30]), Liu and Watbled (2009, [98, 99]), Lacoin (2010, [73]).

We assume only $\mathbb{Q}[|\eta(0,0)|^p] < +\infty$ for some $p > 2d$ (notice that we do not suppose that $\lambda(\beta) < +\infty$, in fact, it is possible that $\lambda(\beta) = +\infty$). We are interested in the asymptotic behaviour of the partition function $Z_n(\beta)$ and the free energy $\frac{1}{n}\ln Z_n(\beta)$. For simplicity, we shall write W_n for $W_n(\beta)$, Z_n for $Z_n(\beta)$ and η for $\eta(0,0)$. We use the same letter η to denote the environment sequence $(\eta(n,x))_{(n,x)\in\mathbb{N}\times\mathbb{Z}^d}$ and the random variable $\eta(0,0)$; there will be no confusion, because of the context.

In the following, we shall prove a concentration inequality for the free energy $\frac{\ln Z_n}{n}$ and a convergence result for the centered energy $\frac{\ln Z_n}{n} - \frac{\mathbb{Q}[\ln Z_n]}{n}$: cf. Theorem 5.1.

Let $\beta > 0$ be fixed such that $\mathbb{Q}[e^{\beta\eta}] < +\infty$ and set $K = 2\exp(\lambda(-\beta) + \lambda(\beta))$, Liu and Watbled (2009, Theorem 6.5 in [98]) proved that

$$\frac{1}{n}\ln W_n - \frac{1}{n}\mathbb{Q}[\ln W_n] \overset{n\to\infty}{\longrightarrow} 0 \quad \mathbb{Q}\text{-a.s. and in } \mathbb{L}^p(\mathbb{Q}) \ (p > 0),$$

with

$$\limsup_{n\to+\infty} \sqrt{\frac{n}{\ln n}}\left|\frac{\ln W_n}{n} - \frac{\mathbb{Q}[\ln W_n]}{n}\right| \le 2\sqrt{K} \quad \text{a.s.,}$$

and for every $p > 0$,

$$\limsup_{n\to+\infty} n^{\frac{p}{2}}\mathbb{Q}\left[\left|\frac{\ln W_n - \mathbb{Q}[\ln W_n]}{n}\right|^p\right] \le p2^p K^{\frac{p}{2}}\Gamma\left(\frac{p}{2}\right).$$

5.1   Main result

Here we first prove that the centered energy $\frac{\ln Z_n}{n} - \frac{\mathbb{Q}[\ln Z_n]}{n}$ converges to 0 in $\mathbb{L}^p$ under the weaker condition that $\mathbb{Q}[|\eta|^p] < +\infty$, for $p > 2d$.

Theorem 5.1 (*Concentration inequality for the free energy*) *If for some $p > 2d$,*

$$\mathbb{Q}\left[|\eta|^p\right] < +\infty, \tag{5.4}$$

then

$$\frac{1}{n}\ln Z_n - \frac{1}{n}\mathbb{Q}[\ln Z_n] \overset{n \to \infty}{\longrightarrow} 0 \quad in \; \mathbb{L}^p(\mathbb{Q}). \tag{5.5}$$

Moreover, if (5.4) *holds for some $p \geq 2$, then*

$$\mathbb{Q}\left[\left|\frac{\ln Z_n - \mathbb{Q}[\ln Z_n]}{n}\right|^p\right] \leq Cn^{-(\frac{p}{2}-d)}. \tag{5.6}$$

where $C > 0$ is a constant depending only on β, p and the law of η.

Let $\beta > 0$ be fixed such that $\mathbb{Q}[e^{\beta\eta}] < +\infty$, Liu and Watbled (2009, Lemma 7.1 in [98]) proved that

$$p_-(\beta) = \lim_{n \to \infty} \frac{1}{n}\ln W_n \in [\beta\mathbb{Q}[\eta] - \lambda(\beta), \, 0] \quad \mathbb{Q}\text{-a.s. and in } \mathbb{L}^p(\mathbb{Q}) \; (p \geq 1),$$

where $p_-(\beta) = \lim_{n \to \infty} \frac{1}{n}\mathbb{Q}[\ln W_n]$.

Suppose that $\int_0^\infty (1 - F(x))^{\frac{1}{d+1}}\,dx < +\infty$, $\int_{-\infty}^0 F(x)^{\frac{1}{d+1}}\,dx < +\infty$ and $\mathbb{Q}|\eta| < +\infty$, Vargas (2007, Theorem 3.1 in [136]) proved that

$$\frac{\ln Z_n}{n} \overset{n \to \infty}{\longrightarrow} p(\beta), \quad \mathbb{Q}\text{-a.s. and in } \mathbb{L}^1(\mathbb{Q}), \tag{5.7}$$

where $p(\beta) = \lim_{n \to \infty} \frac{1}{n}\mathbb{Q}[\ln Z_n]$.

As a consequence of Theorem 5.1 and a supplement of Vargas (2007, Theorem 3.1 in [136]), we obtain a result about convergence in $\mathbb{L}^p$ of the free energy.

Corollary 5.1 (*Convergence in $\mathbb{L}^p$ for the free energy*) *If for some $p > 2d$,* (5.4) *holds, then*

$$\frac{\ln Z_n}{n} \overset{n \to \infty}{\longrightarrow} p(\beta) \quad in \; \mathbb{L}^1(\mathbb{Q}). \tag{5.8}$$

5.2 Proof of the main result

In this section, we shall prove Theorem 5.1.

Proof of Theorem 5.1. As in [28], we write $\ln Z_n - \mathbb{Q}[\ln Z_n]$ as a sum of $(\mathcal{E}_j)_{1 \leq j \leq n}$ martingale differences:

$$\ln Z_n - \mathbb{Q}[\ln Z_n] = \sum_{j=1}^n V_{n,j}, \quad \text{with } V_{n,j} = \mathbb{Q}_j[\ln Z_n] - \mathbb{Q}_{j-1}[\ln Z_n], \tag{5.9}$$

where $\mathbb{Q}_j$ denotes the conditional expectation with respect to $\mathbb{Q}$ given $\mathcal{E}_j$, $\mathcal{E}_0 = \{\emptyset, E\}$ and $\mathcal{E}_j = \sigma\{\eta(i,x) : 1 \leq i \leq j, x \in \mathbb{Z}^d\}$ for $j \geq 1$. Set

$$e_{n,j} = \exp\left(\beta \sum_{1 \leq k \leq n, k \neq j} \eta(k,\omega_k)\right), \quad Z_{n,j} = \mathbb{P}[e_{n,j}]. \tag{5.10}$$

Since $\mathbb{Q}_{j-1}[\ln Z_{n,j}] = \mathbb{Q}_j[\ln Z_{n,j}]$, we have

$$V_{n,j} = \mathbb{Q}_j\left[\ln \frac{Z_n}{Z_{n,j}}\right] - \mathbb{Q}_{j-1}\left[\ln \frac{Z_n}{Z_{n,j}}\right]. \tag{5.11}$$

We denote by $\mathbb{L}_j = \{-j, -j+1, \ldots, j-1, j\}^d$ the location of the simple random walk in time j, $j \in \mathbb{N}$. For $j \in \mathbb{N}$ and $x \in \mathbb{Z}^d$, define

$$\bar{\eta}_x = \bar{\eta}(j,x) = \exp(\beta\eta(j,x)), \quad \alpha_x = \frac{\mathbb{P}[e_{n,j}; \omega_j = x]}{Z_{n,j}}. \tag{5.12}$$

Then

$$\sum_{x \in \mathbb{L}_j} \alpha_x = 1 \text{ and } \frac{Z_n}{Z_{n,j}} = \sum_{x \in \mathbb{L}_j} \alpha_x \bar{\eta}_x. \tag{5.13}$$

By (5.11), using Jensen's inequality (the function $x \mapsto x^p$ $(p > 1)$ is convex), we get

$$\mathbb{Q}_{j-1}\left[|V_{n,j}|^p\right] = \mathbb{Q}_{j-1}\left[\left|\mathbb{Q}_j\left[\ln \frac{Z_n}{Z_{n,j}}\right] - \mathbb{Q}_{j-1}\left[\ln \frac{Z_n}{Z_{n,j}}\right]\right|^p\right]$$

$$\leq 2^{p-1}\left(\mathbb{Q}_{j-1}\left[\left|\mathbb{Q}_j\left[\ln \frac{Z_n}{Z_{n,j}}\right]\right|^p\right] + \mathbb{Q}_{j-1}\left[\left|\mathbb{Q}_{j-1}\left[\ln \frac{Z_n}{Z_{n,j}}\right]\right|^p\right]\right)$$

$$\leq 2^{p-1}\left(\mathbb{Q}_{j-1}\left[\mathbb{Q}_j^p\left[\left|\ln \frac{Z_n}{Z_{n,j}}\right|\right]\right] + \mathbb{Q}_{j-1}\left[\mathbb{Q}_{j-1}^p\left[\left|\ln \frac{Z_n}{Z_{n,j}}\right|\right]\right]\right)$$

$$\leq 2^{p-1}\left(\mathbb{Q}_{j-1}\left[\left|\ln \frac{Z_n}{Z_{n,j}}\right|^p\right] + \mathbb{Q}_{j-1}\left[\left|\ln \frac{Z_n}{Z_{n,j}}\right|^p\right]\right)$$

$$\leq 2^p \mathbb{Q}_{j-1}\left[\left|\ln \frac{Z_n}{Z_{n,j}}\right|^p\right].$$

So we have

$$\mathbb{Q}\left[|V_{n,j}|^p\right] \leq 2^p \mathbb{Q}\left[\left|\ln \frac{Z_n}{Z_{n,j}}\right|^p\right]. \tag{5.14}$$

Define functions

$$f(x) = \ln_-^p x, \ x \in \mathbb{R}^+, \ p \geq 1$$

(where and hereafter $X_- = \max(0, -X)$) and

$$g(x) = \begin{cases} p^p e^{-p} x & \text{if } 0 < x < e^p, \\ \ln^p x & \text{if } x \geq e^p. \end{cases}$$

Then

$$|\ln x|^p \leq f(x) + g(x), x \in \mathbb{R}^+.$$

Since $f(x)$ is convex and $g(x)$ is concave, using Jensen's inequality, we get

$$\mathbb{Q}_j \left[\left| \ln \frac{Z_n}{Z_{n,j}} \right|^p \right] \leq \mathbb{Q}_j \left[f \left(\frac{Z_n}{Z_{n,j}} \right) \right] + \mathbb{Q}_j \left[g \left(\frac{Z_n}{Z_{n,j}} \right) \right]$$

$$= \mathbb{Q}_j \left[f \left(\sum_{x \in \mathbb{L}_j} \alpha_x \overline{\eta}_x \right) \right] + \mathbb{Q}_j \left[g \left(\sum_{x \in \mathbb{L}_j} \alpha_x \overline{\eta}_x \right) \right]$$

$$\leq \mathbb{Q}_j \left[\sum_{x \in \mathbb{L}_j} \alpha_x f(\overline{\eta}_x) \right] + g \left(\sum_{x \in \mathbb{L}_j} \mathbb{Q}_j \left[\alpha_x \overline{\eta}_x \right] \right). \tag{5.15}$$

Since

$$\mathbb{Q}_j \left[\sum_{x \in \mathbb{L}_j} \alpha_x f(\overline{\eta}_x) \right] = \mathbb{Q}_j \left[\sum_{x \in \mathbb{L}_j} \alpha_x f(\exp(\beta \eta(j, x))) \right]$$

$$= \mathbb{Q}_j \left[\sum_{x \in \mathbb{L}_j} \alpha_x \beta^p \eta^p_-(j, x) \right]$$

$$= \beta^p \sum_{x \in \mathbb{L}_j} \mathbb{Q}_j[\alpha_x] \mathbb{Q}_j \left[\eta^p_-(j, x) \right]$$

$$= \beta^p \sum_{x \in \mathbb{L}_j} \mathbb{Q}_j[\alpha_x] \eta^p_-(j, x)$$

$$= \beta^p \sum_{x \in \mathbb{L}_j} \mathbb{Q}_{j-1}[\alpha_x] \eta^p_-(j, x),$$

and since

$$g \left(\sum_{x \in \mathbb{L}_j} \mathbb{Q}_j \left[\alpha_x \overline{\eta}_x \right] \right) = g \left(\sum_{x \in \mathbb{L}_j} \mathbb{Q}_j \left[\alpha_x \exp(\beta \eta(j, x)) \right] \right)$$

$$= g \left(\sum_{x \in \mathbb{L}_j} \mathbb{Q}_j[\alpha_x] \exp(\beta \eta(j, x)) \right)$$

$$\leq g \left(\left(\sum_{x \in \mathbb{L}_j} \mathbb{Q}_j[\alpha_x] \right) \max_{x \in \mathbb{L}_j} \exp(\beta \eta(j, x)) \right)$$

$$= g\left(\max_{x\in \mathbb{L}_j} \exp(\beta \eta(j,x))\right)$$

$$\leq p^p + \ln^p\left(\max_{x\in \mathbb{L}_j} \exp(\beta \eta(j,x))\right)$$

$$\leq p^p + \beta^p \max_{x\in \mathbb{L}_j} \eta_+^p(j,x)$$

(where and hereafter $X_+ = \max(0, X)$), together with (5.15), we see that

$$\mathbb{Q}_j\left[\left|\ln \frac{Z_n}{Z_{n,j}}\right|^p\right] \leq \beta^p \sum_{x\in \mathbb{L}_j} \mathbb{Q}_{j-1}[\alpha_x]\eta_-^p(j,x) + p^p + \beta^p \max_{x\in \mathbb{L}_j} \eta_+^p(j,x).$$

So we have

$$\mathbb{Q}\left[\left|\ln \frac{Z_n}{Z_{n,j}}\right|^p\right] \leq \beta^p \mathbb{Q}\left[\eta_-^p(j,x)\right] + p^p + \beta^p \#\{\mathbb{L}_j\}\mathbb{Q}\left[\eta_+^p(j,x)\right], \qquad (5.16)$$

where and hereafter $\#\{\cdot\}$ denotes the cardinality of the set $\{\cdot\}$. Since (5.14) and (5.16), we know that

$$\mathbb{Q}\left[|V_{n,j}|^p\right] \leq 2^p \beta^p \mathbb{Q}\left[\eta_-^p(j,x)\right] + 2^p p^p + 2^p \beta^p \#\{\mathbb{L}_j\}\mathbb{Q}\left[\eta_+^p(j,x)\right]$$

$$\leq 2^p \beta^p \mathbb{Q}\left[\eta_-^p(j,x)\right] + 2^p p^p + 2^p \beta^p (2j)^d \mathbb{Q}\left[\eta_+^p(j,x)\right]. \qquad (5.17)$$

Using Burkholder's inequality, we have that, for $p > 1$,

$$\mathbb{Q}\left[\left|\sum_{j=1}^n V_{n,j}\right|^p\right] \leq B_p^p \mathbb{Q}\left[\left|\sum_{j=1}^n V_{n,j}^2\right|^{p/2}\right], \qquad (5.18)$$

where $B_p = \frac{18p^{3/2}}{(p-1)^{1/2}}$. Using Holder's inequality, we obtain that, for $p > 2$,

$$\sum_{j=1}^n V_{n,j}^2 \leq n^{1-\frac{2}{p}}\left(\sum_{j=1}^n |V_{n,j}|^p\right)^{2/p}. \qquad (5.19)$$

Obviously, we have that, for $p = 2$,

$$\sum_{j=1}^n V_{n,j}^2 = n^{1-\frac{2}{p}}\left(\sum_{j=1}^n |V_{n,j}|^p\right)^{2/p}. $$

Thus we obtain that (5.19) holds for $p \geq 2$. Since (5.9), (5.18), (5.19) and (5.17) hold for $p \geq 2$, we see that, for $p \geq 2$,

$$\mathbb{Q}\left[\left|\frac{\ln Z_n - \mathbb{Q}[\ln Z_n]}{n}\right|^p\right]$$

$$\leq B_p^p n^{-p} \mathbb{Q}\left[\left|n^{1-\frac{2}{p}}\left(\sum_{j=1}^n |V_{n,j}|^p\right)^{2/p}\right|^{p/2}\right]$$

$$= B_p^p n^{-\frac{p}{2}-1} \sum_{j=1}^n \mathbb{Q}\left[|V_{n,j}|^p\right]$$

$$\leq B_p^p n^{-\frac{p}{2}-1} \sum_{j=1}^n \left[2^p \beta^p \mathbb{Q}\left[\eta_-^p(j,x)\right] + 2^p p^p + 2^p \beta^p (2j)^d \mathbb{Q}\left[\eta_+^p(j,x)\right]\right]$$

$$\leq B_p^p n^{-\frac{p}{2}-1} \sum_{j=1}^n \left[2^p \beta^p \mathbb{Q}\left[|\eta|^p\right] + 2^p p^p + 2^p \beta^p (2j)^d \mathbb{Q}\left[|\eta|^p\right]\right]$$

$$\leq C n^{-\left(\frac{p}{2}-d\right)},$$

where $B_p = \frac{18 p^{3/2}}{(p-1)^{1/2}}$, $C > 0$ is a constant depending only on β, p and the law of η. Thus (5.6) holds. Then (5.5) holds if $p > 2d$. ∎

5.3   Proof of Corollary 5.1

We now prove Corollary 5.1 by use of Theorem 5.1.

Proof of Corollary 5.1. Since

$$\left\|\frac{\ln Z_n}{n} - p(\beta)\right\|_p \leq \left\|\frac{\ln Z_n}{n} - \frac{\mathbb{Q}[\ln Z_n]}{n}\right\|_p + \left\|\frac{\mathbb{Q}[\ln Z_n]}{n} - p(\beta)\right\|_p$$

and

$$\left\|\frac{\mathbb{Q}[\ln Z_n]}{n} - p(\beta)\right\|_p = \left|\frac{\mathbb{Q}[\ln Z_n]}{n} - p(\beta)\right|,$$

we know that (5.8) holds by Theorem 5.1. ∎

Bibliography

[1] Adler A, Rosalsky A, Taylor R L. Strong laws of large numbers for weighted sums of random elements in normed linear spaces[J]. Int. J. Math. Math. Sci., 1989, 12(3): 507-529.

[2] Adler A, Rosalsky A, Taylor R L. Some strong laws of large numbers for sums of random elements[J]. Bull. Inst. Math. Acad. Sin., 1992, 20(4): 335-357.

[3] Agresti A. Bounds on the extinction time distribution of a branching process[J]. Adv. Appl. Prob., 1974, 6(2): 322-335.

[4] Ahmed S E, Antonini R G, Volodin A I. On the rate of complete convergence for weighted sums of arrays of Banach space valued random elements with application to moving average processes[J]. Statist. Probab. Lett., 2002, 58(2): 185-194.

[5] Albeverio S, Zhou X Y. A martingale approach to directed polymers in a random environment[J]. J. Theoret. Probab., 1996, 9(1): 171-189.

[6] Alsmeyer G. Convergence rates in the law of large numbers for martingales[J]. Stochastic Process. Appl., 1990, 36(2): 181-194.

[7] Assouad P. Espaces p-lisses et q-convexes, inégalités de Burkholder[R]. Séminaire Maurey-Schwartz 1974-1975: Espaces $\mathbb{L}^p$, applications radonifiantes et géométrie des espaces de Banach, Exp. No. XV, 8 pp. Centre Math., École Polytech., Paris, 1975.

[8] Athreya K B, Karlin S. Branching processes with random environments I: Extinction Probabilities[J]. Ann. Math. Statis., 1971a, 42: 1499-1520.

[9] Athreya K B, Karlin S. Branching processes with random environments II: Limit Theorems[J]. Ann. Math. Statis., 1971b, 42: 1843-1858.

[10] Athreya K B, Ney P E. Branching processes[M]. New York: Springer, 1972.

[11] Bai Z D, Su C. The complete convergence for partial sums of iid random variables[J]. Sci. China Ser. A, 1985, 28(12): 1261-1277.

[12] Barral J. Moments, continuité, et analyse multifractale des martingales de Mandelbrot[J]. Probab. Theory Related Fields, 1999, 113: 535-569.

[13] Barral J, Jin X, Mandelbrot B. Uniform convergence for complex $[0, 1]$-martingales[J]. Ann. Appl. Prob., 2010, 20(4): 1205-1218.

[14] Barral J, Jin X, Mandelbrot B. Convergence of complex multiplicative cascades[J]. Ann. Appl. Prob., 2010, 20(4): 1219-1252.

[15] Baum L E, Katz M. Convergence rates in the law of large numbers[J]. Trans. Amer. Math. Soc., 1965, 120: 108-123.

[16] Biggins J D, Kyprianou A E. Measure change in multitype branching[J]. Adv. Appl. Prob., 2004, 36(2): 544-581.

[17] Bingham N H, Goldie C M, Teugels J L. Regular variation[M]. Cambridge: Cambridge University Press, 1987.

[18] Birkner M. A condition for weak disorder for directed polymers in random environment[J]. Elect. Comm. in Probab., 2004, 9: 22-25.

[19] Bolthausen E. A note on the diffusion of directed polymers in a random environment[J]. Comm. Math. Phys., 1989, 123(4): 529-534.

[20] Breiman L. Probability[M]. Classics in Applied Mathematics 7, SIAM, 1992.

[21] Carmona P, Guerra F, Hu Y Y, Méjane O. Strong disorder for a certain class of directed polymers in a random environment[J]. J. Theoret. Probab., 2006, 19(1): 134-151.

[22] Carmona P, Hu Y Y, On the partition function of a directed polymers in a Gaussian random environment[J]. Probab. Theory Relat. Fields, 2002, 124: 431-457.

[23] Carmona P, Hu Y Y, Fluctuation exponents and large deviations for directed polymers in a random environment[J]. Stochastic Process. Appl., 2004, 112(2): 285-308.

[24] Chen P, Sung S H, Volodin A I. Rate of complete convergence for arrays of Banach space valued random elements[J]. Siberian Adv. Math., 2006, 16(3): 1-14.

[25] Chow Y S. Delayed sums and Borel summability of independent, identically distributed random variables[J]. Bull. Inst. Math. Acad. Sinica, 1973, 1(2): 207-220.

[26] Chow Y S. On the rate of moment convergence of sample sums and extremes[J]. Bull. Inst. Math. Acad. Sin., 1988, 16: 177-201.

[27] Collet P, Koukiou F. Large deviations for multiplicative chaos[J]. Commun. Math. Phys., 1992, 147: 329-342.

[28] Comets F, Shiga T, Yoshida N. Directed polymers in random environment: Path localization and strong disorder[J]. Bernoulli, 2003, 9(4): 705-723.

[29] Comets F, Shiga T, Yoshida N. Probabilistic analysis of directed polymers in a random environment: a review[C]. Stochastic analysis on large scale interacting systems, 115 – 142, Adv. Stud. Pure Math., 39(4), Math. Soc. Japan, Tokyo, 2004.

[30] Comets F, Vargas V. Majorizing multiplicative cascades for directed polymers in random media[J]. Alea, 2006, 2: 267-277.

[31] Comets F, Yoshida N. Directed polymers in random environment are diffusive at weak disorder[J]. Ann. Probab., 2006, 34(5): 1746-1770.

[32] Dedecker J, Merlevède F. Convergence rates in the law of large numbers for Banach-valued dependent variables[J]. Theory Probab. Appl., 2008, 52(3): 416-438.

[33] Duquesne T. An elementary proof of Hawkes's conjecture on Galton-Watson trees[J]. Electron. Commun. Probab., 2009, 14: 151-164.

[34] Durrett R, Liggett T. Fixed points of the smoothing transformation[J]. Z. Wahrsch. verw. Gebeite, 1983, 64: 275-301.

[35] Erdös P. On a theorem of Hsu and Robbins[J]. Ann. Math. Statist., 1949, 20(2): 286-291.

[36] Gafurov M U, Slastnikov A D. Some problems of the exit of a random walk beyond a curvilinear boundary and large deviations[J]. Theory Probab. Appl., 1987, 32(2): 299-321.

[37] Ghosal S, Chandra T K. Complete convergence of martingale arrays[J]. J. Theoret. Probab., 1998, 11(3): 621-631.

[38] Guivarc'h Y. Sur une extension de la notion de loi semi-stable[J]. Ann. Inst. Henri Poincaré Probab. Statist., 1990, 26(2): 261-285.

[39] Gut A. Complete convergence and Cesàro summation for i.i.d. random variables[J]. Probab. Theory Related Fields, 1993, 97(1-2): 169-178.

[40] Hall P, Heyde C C. Martingale limit theory and its application[M]. New York, USA: Academic Press, 1980.

[41] Hambly B. On the limiting distribution of a supercritical branching process in a random environment[J]. J. Appl. Probab., 1992, 29(3): 499-518.

[42] Hao S L. Convergence rates in the law of large numbers for arrays of Banach valued martingale differences[J]. Abstr. Appl. Anal., 2013, 2013: 1-26.

[43] Hao S L, Liu Q S. Théorèmes de type Baum-Katz pour un tableau de martingales[J]. C. R. Acad. Sci. Paris, Ser. I, 2012, 350(1-2): 91-96.

[44] Hao S L, Liu Q S. Convergence rates in the law of large numbers for arrays of martingale differences[J]. J. Math. Anal. Appl., 2014, 417(2): 733-773.

[45] Hawkes J. Trees generated by a simply branching processes[J]. J. London Math. Soc., 1981, 24: 373-384.

[46] Henley C L, Huse D A. Pinning and roughening of domain walls in ising systems due to random impurities[J]. Phys. Rev. Lett., 1985, 54: 2708-2711.

[47] Hernández V, Urmeneta H, Volodin A. On complete convergence for arrays of random elements and variables[J]. Stochastic Anal. Appl., 2007, 25(2): 281-291.

[48] Holley R, Waymire E C. Multifractal dimensions and scaling exponents for strongly bounded cascades[J]. Ann. Appl. Prob., 1992, 2: 819-845.

[49] Hsu P L, Robbins H. Complete convergence and the law of large numbers[J]. Proc. Natl. Acad. Sci. USA, 1947, 33(2): 25-31.

[50] Hu T C, Cabrera M O, Sung S H, Volodin A, Complete convergence for arrays of rowwise independent random variables[J]. Commun. Korean Math. Soc., 2003, 18: 375-383.

[51] Hu T C, Chang H C. Stability for randomly weighted sums of random elements[J]. Internat. J. Math. Math. Sci., 1999, 22(3): 559-568.

[52] Hu T C, Li D L, Rosalsky A, Volodin A. On the rate of complete convergence for weighted sums of Banach spaces valued random elements[J]. Theory Probab. Appl., 2003, 47(3): 455-468.

[53] Hu T C, Móricz F, Taylor R L. Strong laws of large numbers for arrays of rowwise independent random variables[J]. Acta Math. Hungar., 1989, 54(1-2): 153-162.

[54] Hu T C, Ordóñez Cabrera M, Sung S H, Volodin A I. Complete convergence for arrays of rowwise independent random variables[J]. Commun. Korean Math. Soc., 2003, 18(2): 375-383.

[55] Hu T C, Ordóñez Cabrera M, Volodin A I. Convergence of randomly weighted sums of Banach space valued random elements and uniform integrability concerning the random weights[J]. Statist. Probab. Lett., 2001, 51(2): 155-164.

[56] Hu T C, Rosalsky A, Szynal D, Volodin A I. On complete convergence for arrays of rowwise independent random elements in Banach spaces[J]. Stochastic Anal. Appl., 1999, 17(6): 963-992.

[57] Hu T C, Szynal D, Volodin A I. A note on complete convergence for arrays[J]. Statist. Probab. Lett., 1998, 38(1): 27-31.

[58] Hu T C, Volodin A I. Addendum to "A note on complete convergence for array", Statist. Probab. Lett. 38 (1998), no. 1, 27-31[J]. Statist. Probab. Lett., 2000, 47(2): 209-211.

[59] Huang C M. Théorèmes limites et vitesses de convergence pour certains processus de branchement et des marches aléatoires branchantes[D]. Vannes: Univ. Bretagne-Sud, 2010.

[60] Huang C M, Liu Q S. Moments, moderate and large deviations for a branching process in a random environment[J]. Stochastic Process. Appl., 2012, 122: 522-545.

[61] Imbrie J Z, Spencer T. Diffusion of directed polymers in a random environment[J]. J. Statist. Phys., 1988, 52(3-4): 609-626.

[62] Jain N C. Tail probabilities for sums of independent Banach space valued random variables[J]. Z. Wahrsh. Verw. Gebiete., 1975, 33(3): 155-166.

[63] Joffe A. Remarks on the structure of trees with applications to supercritical Galton-Watson processes[J]. in: Joffe A, Ney P. (Eds.), Advances in Prob., Dekker, New York, 1978, 5: 263-268.

[64] Kahane J P, Peyrière J. Sur certaines martingales de Benoit Mandelbrot[J]. Adv. Math., 1976, 22: 131-145.

[65] Karamata J. Sur un mode de croissance régulière des fonctions[J]. Mathematica (Cluj), 1930, 4: 38-53.

[66] Katz M. The probability in the tail of a distribution[J]. Ann. Math. Statist., 1963, 34(1): 312-318.

[67] Kifer Y. The Burgers equation with a random force and a general model for directed polymers in random environment[J]. Probab. Theory Relat. Fields, 1997, 108: 29-65.

[68] Kinnison A L, Mörters P. Simultaneous multifractal analysis of the branching and visibility measure on a Galton-Watson tree[J]. Adv. in Appl. Probab., 2010, 42(1): 226-245.

[69] Korevaar J, van Aardenne-Ehrenfest T, de Bruijn N G. A note on slowly oscillating functions[J]. Nieuw Arch. Wiskd., 1949, 23(2): 77-86.

[70] Kruglov V M, Volodin A I, Hu T C. On complete convergence for arrays[J]. Statist. Probab. Lett., 2006, 76(15): 1631-1640.

[71] Kuczmaszewska A. On some conditions for complete convergence for arrays[J]. Statist. Probab. Lett., 2004, 66(4): 399-405.

[72] Kuhlbusch D. On weighted branching processes in random environment[J]. Stochastic Process. Appl., 2004, 109(1): 113-144.

[73] Lacoin H. New bounds for the free energy of directed polymers in dimension 1+1 and 1+2[J]. Commun. Math. Phys., 2010, 294: 471-503.

[74] Laha R G, Rohatgi V K. Probability Theory[M]. John Wiley & Sons, Inc., 1979.

[75] Lai T L. Limit theorems for delayed sums[J]. Ann. Probab., 1974, 2(3): 432-440.

[76] Lanzinger H, Stadtmüller U. Baum-Katz laws for certain weighted sums of independent and identically distributed random variables[J]. Bernoulli, 2003, 9(6): 985-1002.

[77] Ledoux M. Concentration of measure and logarithmic Sobolev inequalities[R]. In Séminaire de Probabilités, XXXIII, volume 1709 of Lecture Notes in Math., 120-216. Berlin: Springer, 1999.

[78] Lesigne E, Volný D. Large deviations for martingales[J]. Stochastic Process. Appl., 2001, 96(1): 143-159.

[79] Li D L, Rao M B, Jiang T F, Wang X C. Complete convergence and almost sure convergence of weighted sums of random variables[J]. J. Theoret. Probab., 1995, 8(1): 49-76.

[80] Li D L, Zhang F X, Rosalsky A. A Supplement to the Baum-Katz-Spitzer Complete Convergence Theorem[J]. Acta. Math. Sin. (Engl. Ser.), 2007, 23(3): 557-562.

[81] Liang H Y. Complete convergence for weighted sums of negatively associated random variables[J]. Statist. Probab. Lett., 2000, 48: 317-325.

[82] Li D L, Zhang F X, Rosalsky A. A Supplement to the Baum-Katz-Spitzer Complete Convergence Theorem[J]. Acta. Math. Sin. (Engl. Ser.), 2007, 23(3): 557-562.

[83] Li Y Q, Hu Y L, Liu Q S. Weighted moments for a supercritical branching process in a varying or random environment[J]. Sci China Math, 2011, 54(7): 1437 - 1444.

[84] Liang H Y. Some discussion on the conditions of complete convergence for sums of B-valued random elements[J]. J. Systems Sci. Math. Sci., 2000, 13(2): 164-169.

[85] Liang H Y, Wang L. Convergence rates in the law of large numbers for B-valued random elements[J]. Acta. Math. Sci. Ser. B Engl. Ed., 2001, 21(2): 229-236.

[86] Liang X G. Propriétés asymptotiques des martingales de Mandelbrot et des marches aléatoires branchantes[D]. Vannes: Univ. Bretagne-Sud, 2010.

[87] Liu Q S. The exact Hausdorff dimension of a branching set[J]. Probab. Theory Related Fields., 1996, 104: 515-538.

[88] Liu Q S. Sur une équation fonctionelle et ses applications: une extension du théorème de Kesten-Stigum concernant des processus de branchement[J]. Adv. Appl. Probab., 1997, 29: 353-373.

[89] Liu Q S. Fixed points of a generalized smoothing transformation and applications to the branching random walk[J]. Adv. Appl. Prob., 1998, 30: 85-112.

[90] Liu Q S. Exact packing measure of boundary of a Galton-Watson tree[J]. Stoch. Proc. Appl., 2000, 85: 19-28.

[91] Liu Q S. On generalized multiplicative cascades[J]. Stoch. Proc. Appl., 2000, 86: 263-286.

[92] Liu Q S. Asymptotic properties and absolute continuity of laws stable by random weighted mean[J]. Stoch. Proc. Appl., 2001, 95: 83-107.

[93] Liu Q S. Local dimension of the branching measure on a Galton-Watson tree[J]. Ann. Inst. H. Poincare Probab. Statist., 2001, 37: 195-222.

[94] Liu Q S, Rio E, Rouault A. Limit theorem for multiplicative processes[J]. J. Theoret. Probab., 2003, 16: 971-1014.

[95] Liu Q S, Rouault A. On two measures defined on the boundary of a branching tree[C]. in: Athreya K B, Jagers P. (Eds.), Classical and Modern Branching Processes, IMA Volumes in Mathematics and its Applications, Springer-Verlag, 1996(84): 187-202.

[96] Liu Q S. Rouault A. Limit theorem for Mandelbrot's multiplicative cascades[J]. Ann. Appl. Prob., 2000, 10: 218-231.

[97] Liu Q S, Shieh N. R. A uniform limit law for the branching measure on a Galton-Watson tree[J]. Asian J. Math., 1999, 3: 381-386.

[98] Liu Q S, Watbled F. Exponential inequalities for martingales and asymptotic properties of the free energy of directed polymers in a random environment[J]. Stochastic Process. Appl., 2009, 119(10): 3101-3132.

[99] Liu Q S, Watbled F. Large deviation inequalities for supermartingales and applications to directed polymers in a random environment[J]. C. R. Acad. Sci. Paris, Ser. I, 2009, 347: 1207-1212.

[100] Lyons R, Pemantle R, Peres Y. Ergodic theorey on Galton-Watson trees, Speed of random walk and dimension of harmonic measure[J]. Ergodic Theory Dynamical Systems., 1995, 15: 593-619.

[101] Mandelbrot B. Multiplications aléatoires et distributions invariantes par moyenne pondérée aléatoire[J]. C. R. Acad. Sci. Paris Ser. A, 1974, 278: 289-292; 355-358.

[102] Mandelbrot B. Intermittent turbulence in self-similar cascades: divergence of high moments and dimension of the carrier[J]. J. Fluid Mech., 1974, 62: 331-333.

[103] Martin J B. Limiting shape for directed percolation models[J]. Ann. Probab., 2004, 32(4): 2908-2937.

[104] Matuszewska W. A remark on my paper "Regularly increasing functions in the theory of $\mathbb{L}^{*\phi}$-spaces"[J]. Studia Math., 1965, 25: 265-269.

[105] Mejane O. Upper bound of a volume exponent for directed polymers in a random environment[J]. Ann. I. H. Poincaré-PR, 2004, 40: 299-308.

[106] Menshikov M, Petritis D, Popov S. A note on matrix multiplicative cascades and bindweeds[J]. Markov Process. Related Fields, 2005, 11(1): 37-54.

[107] Neveu J. Arbre et processus de Galton-Watson[J]. Ann. Inst. Henri Poincaré, 1986, 22: 199-207.

[108] O'Brien G L. A limit theorem for sample maxima and heavy branches in Galton-Watson trees[J]. J. Appl. Probab., 1980, 17: 539-545.

[109] Ordóñez Cabrera M. Limit theorems for randomly weighted sums of random elements in normed linear spaces[J]. J. Multivariate Anal., 1988, 25(1): 139-145.

[110] Pisier G. Martingales with values in uniformly convex spaces[J]. Israel J. Math., 1975, 20(3-4): 326-350.

[111] Potter H S A. On the mean value of a Dirichlet series II[J]. Proc. Lond. Math. Soc., 1942, 47(2): 1-19.

[112] Qiu D H, Hu T C, Cabrera M O, Volodin A. Complete convergence for weighted sums of arrays of banach-space-valued random elements[J]. Lith. Math. J., 2012, 52(3): 316-325.

[113] Ramachandran B. On the order and the type of entire characteristic functions[J]. Ann. Stat., 1962, 33: 1238-1255.

[114] Rosalsky A, Sreehari M. On the limiting behavior of randomly weighted partial sums[J]. Statist. Probab. Lett., 1998, 40(4): 403-410.

[115] Shieh N-R, Taylor S J. Multifractal spectra of branching measure on a Galton-Watson tree[J]. J. Appl. Probab., 2002, 39(1): 100-111.

[116] Spitzer F. A combinatorial lemma and its applications to probability theory[J]. Trans. Amer. Math. Soc., 1956, 82(2): 323-339.

[117] Stoica G. A note on the rate of convergence in the strong law of large numbers for martingales[J]. J. Math. Anal. Appl. 2011, 381(2): 910-913.

[118] Stout W F. Some results on the complete and almost sure convergence of linear combinations of independent random variables and martingale differences[J]. Ann. Math. Statist., 1968, 39(5): 1749-1562.

[119] Sung S H. Complete convergence for weighted sums of arrays of rowwise independent B-valued random variables[J]. Stochastic Anal. Appl., 1997, 15(2): 255-267.

[120] Sung S H. Moment inequalities and complete moment convergence[J]. J. Inequal. Appl., 2009, 2009: 1-14.

[121] Sung S H. Complete convergence for weighted sums of random elements[J]. Bull. Korean Math. Soc., 2010, 47(2): 369-383.

[122] Sung S H, Volodin A I. On the rate of complete convergence for weighted sums of arrays of random elements[J]. J. Korean Math. Soc., 2006, 43(4): 815-828.

[123] Sung S H, Volodin A I. A note on the rate of complete convergence for weighted sums of arrays of Banach space valued random elements[J]. Stochastic Anal. Appl., 2011, 29(2): 282-291.

[124] Sung S H, Volodin A I, Hu T C. More on complete convergence for arrays[J]. Statist. Probab. Lett., 2005, 71(4): 303-311.

[125] Talagrand M. Concentration of measure and isoperimetric inequalities in product spaces[J]. Inst. Hautes Études Sci. Publ. Math., 1995, 81: 73-205.

[126] Talagrand M. New concentration inequalities in product spaces[J]. Invent. Math., 1996, 126(3): 505-563.

[127] Taylor R L. Weak laws of large numbers in normed linear spaces[J]. Ann. Math. Statist., 1972, 43(4): 1267-1274.

[128] Taylor R L, Calhoun C A. On the almost sure convergence of randomly weighted sums of random elements[J]. Ann. Probab., 1983, 11(3): 795-797.

[129] Taylor R L, Padgett W J. Some laws of large numbers for normed linear spaces[J]. Sankhya Ser. A, 1974, 36(4): 359-368.

[130] Taylor R L, Padgett W J. Weak laws of large numbers in Banach spaces and their extensions[J]. Lecture Notes in Mathematics, Springer, Berlin, 1976, 526: 227-242.

[131] Taylor R L, Raina C C, Daffer P Z. Stochastic convergence of randomly weighted sums of random elements[J]. Stochastic Anal. Appl., 1984, 2(3): 299-321.

[132] Teicher H. Some new conditions for the strong law[J]. Proc. Nat. Acad. Soc., 1968, 59: 705-707.

[133] Thanh L V, Yin G. Almost sure and complete convergence of randomly weighted sums of independent random elements in Banach space[J]. Taiwanese J. Math., 2011, 15(4): 1759-1781.

[134] Thanh L V, Yin G, Wang L Y. State observers with random sampling times and convergence analysis of double-indexed and randomly weighted sums of mixing processes[J]. SIAM J. Control Optim., 2011, 49(1): 106-124.

[135] Tómács T. Convergence rates in the law of large numbers for arrays of Banach space valued random elements[J]. Statist. Probab. Lett., 2005, 72(1): 59-69.

[136] Vargas V. Strong localization and macroscopic atoms for directed polymers[J]. Probab. Theory Relat. Fields, 2007, 134(3/4): 391-410.

[137] Wang F Y. Functional inequalities, Markov semigroups and spectral theory[M]. Science Press, Beijing, New York, 2005.

[138] Wang L Y, Li C, Yin G, Guo L, Xu C Z. State observability and observers of linear-time-invariant systems under irregular sampling and sensor limitations[J]. IEEE T. Automat. Contr., 2011, 56(1): 2639-2654.

[139] Wang X, Rao M B. Convergence in the r-th mean and some weak laws of large numbers for random weighted sums of random elements in Banach spaces[J]. Northeast. Math. J., 1995, 11(1): 113-126.

[140] Wang X C, Rao M B, Yang X Y. Convergence rates on strong laws of large numbers for arrays of rowwise independent elements[J]. Stochastic Anal. Appl., 1993, 11(1): 115-132.

[141] Wang X J, Hu S H. Complete convergence and complete moment convergence for martingale difference sequences[J]. Acta Math. Sin. (Engl. Ser.), 2014, 30(1): 119-132.

[142] Wang X J, Hu S H, Yang W Z, Wang X H. Convergence rates in the strong law of large numbers for martingale differences sequences[J]. Abstr. Appl. Anal., 2012, 2012: 1-13.

[143] Wang Y B, Liu X G, Su C. Equivalent conditions of complete convergence for independent weighted sums[J]. Sci. China Ser. A, 1998, 41(9): 939-949.

[144] Waymire E C, Williams D. A cascades decomposition theory with applications to Markov and exchangeable cascades[J]. Trans. Amer. Math. Soc., 1996, 348: 585-632.

[145] Wei D, Taylor R L. Geometrical consideration of weighted sums convergence and random weighting[J]. Bull. Inst. Math. Acad. Sinica, 1978, 6(1): 49-59.

[146] Wei D, Taylor R L. Convergence of weighted sums of tight random elements[J]. J. Multivariate Anal., 1978, 8(2): 282-294.

[147] Yang W Z, Wang Y W, Wang X H, Hu S H. Complete moment convergence for randomly weighted sums of martingale differences[J]. J. Inequal. Appl., 2013, 2013: 1-13.

[148] Yang X Y, Wang X C. Tail probabilities for sums of independent and identically distributed Banach space valued random elements[J]. Northeast. Math. J., 1986, 2(3): 327 - 338.

[149] Yang W Z, Wang X H, Li X Q, Hu S H. The convergence of double-indexed weighted sums of martingale differences and its application[J]. Abstr. Appl. Anal., 2014, 2014: 1-7.

[150] Yu K F. Complete convergence of weighted sums of martingale differences[J]. J. Theoret. Probab., 1990, 3(2): 339-347.

Index